Learn

Eureka Math™
Grade 2
Modules 4 & 5

Learn ✦ Practice ✦ Succeed

Eureka Math™ student materials for *A Story of Units*® (K–5) are available in the *Learn, Practice, Succeed* trio. This series supports differentiation and remediation while keeping student materials organized and accessible. Educators will find that the *Learn, Practice,* and *Succeed* series also offers coherent—and therefore, more effective—resources for Response to Intervention (RTI), extra practice, and summer learning.

Learn

Eureka Math Learn serves as a student's in-class companion where they show their thinking, share what they know, and watch their knowledge build every day. *Learn* assembles the daily classwork—Application Problems, Exit Tickets, Problem Sets, templates—in an easily stored and navigated volume.

Practice

Each *Eureka Math* lesson begins with a series of energetic, joyous fluency activities, including those found in *Eureka Math Practice*. Students who are fluent in their math facts can master more material more deeply. With *Practice*, students build competence in newly acquired skills and reinforce previous learning in preparation for the next lesson.

Together, *Learn* and *Practice* provide all the print materials students will use for their core math instruction.

Succeed

Eureka Math Succeed enables students to work individually toward mastery. These additional problem sets align lesson by lesson with classroom instruction, making them ideal for use as homework or extra practice. Each problem set is accompanied by a Homework Helper, a set of worked examples that illustrate how to solve similar problems.

Teachers and tutors can use *Succeed* books from prior grade levels as curriculum-consistent tools for filling gaps in foundational knowledge. Students will thrive and progress more quickly as familiar models facilitate connections to their current grade-level content.

Students, families, and educators:

Thank you for being part of the *Eureka Math*™ community, where we celebrate the joy, wonder, and thrill of mathematics.

In the *Eureka Math* classroom, new learning is activated through rich experiences and dialogue. The *Learn* book puts in each student's hands the prompts and problem sequences they need to express and consolidate their learning in class.

What is in the Learn *book?*

Application Problems: Problem solving in a real-world context is a daily part of *Eureka Math*. Students build confidence and perseverance as they apply their knowledge in new and varied situations. The curriculum encourages students to use the RDW process—Read the problem, Draw to make sense of the problem, and Write an equation and a solution. Teachers facilitate as students share their work and explain their solution strategies to one another.

Problem Sets: A carefully sequenced Problem Set provides an in-class opportunity for independent work, with multiple entry points for differentiation. Teachers can use the Preparation and Customization process to select "Must Do" problems for each student. Some students will complete more problems than others; what is important is that all students have a 10-minute period to immediately exercise what they've learned, with light support from their teacher.

Students bring the Problem Set with them to the culminating point of each lesson: the Student Debrief. Here, students reflect with their peers and their teacher, articulating and consolidating what they wondered, noticed, and learned that day.

Exit Tickets: Students show their teacher what they know through their work on the daily Exit Ticket. This check for understanding provides the teacher with valuable real-time evidence of the efficacy of that day's instruction, giving critical insight into where to focus next.

Templates: From time to time, the Application Problem, Problem Set, or other classroom activity requires that students have their own copy of a picture, reusable model, or data set. Each of these templates is provided with the first lesson that requires it.

Where can I learn more about Eureka Math *resources?*

The Great Minds® team is committed to supporting students, families, and educators with an ever-growing library of resources, available at eureka-math.org. The website also offers inspiring stories of success in the *Eureka Math* community. Share your insights and accomplishments with fellow users by becoming a *Eureka Math* Champion.

Best wishes for a year filled with aha moments!

Jill Diniz

Jill Diniz
Director of Mathematics
Great Minds

The Read–Draw–Write Process

The *Eureka Math* curriculum supports students as they problem-solve by using a simple, repeatable process introduced by the teacher. The Read–Draw–Write (RDW) process calls for students to

1. Read the problem.

2. Draw and label.

3. Write an equation.

4. Write a word sentence (statement).

Educators are encouraged to scaffold the process by interjecting questions such as

- What do you see?

- Can you draw something?

- What conclusions can you make from your drawing?

The more students participate in reasoning through problems with this systematic, open approach, the more they internalize the thought process and apply it instinctively for years to come.

Contents

Module 4: Addition and Subtraction Within 200 with Word Problems to 100

Topic E: Strategies for Decomposing Tens and Hundreds

Topic F: Student Explanations of Written Methods

Module 5: Addition and Subtraction Within 1,000 with Word Problems to 100

Grade 2
Module 4

R (Read the problem carefully.)

In the morning, Jacob found 23 seashells on the beach. In the afternoon, he found 10 more. In the evening, he found 1 more.

How many seashells did Jacob find in all? If he gives 10 to his brother, how many seashells will Jacob have left?

D (Draw a picture.)

W (Write and solve an equation.)

Lesson 1:   Relate 1 more, 1 less, 10 more, and 10 less to addition and subtraction of 1 and 10.

©2018 Great Minds®. eureka-math.org

3

W (Write a statement that matches the story.)

_____

_____

_____

Lesson 1: Relate 1 more, 1 less, 10 more, and 10 less to addition and subtraction of 1 and 10.

©2018 Great Minds®. eureka-math.org

EUREKA MATH™

Name _____   Date _____

1. Complete each *more* or *less* statement.

    a. 1 more than 66 is _____67_____.

    b. 10 more than 66 is _____76_____.

    c. 1 less than 66 is _____65_____.

    d. 10 less than 66 is _____56_____.

    e. 56 is 10 more than _____46_____.

    f. 88 is 1 less than _____87_____.

    g. _____57_____ is 10 less than 67.

    h. _____73_____ is 1 more than 72.

    i. 86 is _____10 less_____ than 96.

    j. 78 is _____1 more_____ than 79.

2. Circle the rule for each pattern.

    a. 34, 33, 32, 31, 30, 29   (1 less)   1 more   10 less   10 more

    b. 53, 63, 73, 83, 93   1 less   1 more   10 less   (10 more)

3. Complete each pattern.

    a. 37, 38, 39, __40__, __41__, __42__

    b. 68, 58, 48, __38__, __28__, __18__

    c. 51, 50, __49__, __48__, __47__, 46

    d. 9, 19, __29__, __39__, __49__, 59

EUREKA MATH   Lesson 1:   Relate 1 more, 1 less, 10 more, and 10 less to addition and subtraction   5
                                   of 1 and 10.

©2018 Great Minds®. eureka-math.org

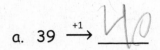

4. Complete each statement to show mental math using the arrow way.

   a. $39 \xrightarrow{+1}$ ___40___ $56 \xrightarrow{+10}$ ___60___ $42 \xrightarrow{-10}$ ___52___ $80 \xrightarrow{-1}$ ___81___

   b. $32 \xrightarrow{+1}$ ___33___ $\xrightarrow{+\_\_}$ 43 $87 \xrightarrow{-10}$ ___70___ $\xrightarrow{-1}$ ___77___

   c. $48 \xrightarrow{+10}$ _____ $\xrightarrow{+\_\_}$ 68 $\xrightarrow{+10}$ ___80___ $\xrightarrow{+1}$ ___80___ $\xrightarrow{+1}$ ___81___

5. Complete each sequence.

   a. $45 \xrightarrow{+10}$ ___70___ $\xrightarrow{-1}$ ___60___ $\xrightarrow{-1}$ ___70___ $\xrightarrow{-10}$ ___80___ $\xrightarrow{-10}$ ___90___

   b. $61 \xrightarrow{-1}$ ___60___ $\xrightarrow{-1}$ ___63___ $\xrightarrow{+10}$ ___70___ $\xrightarrow{+10}$ ___67___ $\xrightarrow{-1}$ ___60___

6. Solve each word problem using the arrow way to record your mental math.

   a. Yesterday Isaiah made 39 favor bags for his party. Today he made 23 more. How many favor bags did he make for his party?

   b. There are 61 balloons. 12 blew away. How many are left?

6

Lesson 1:   Relate 1 more, 1 less, 10 more, and 10 less to addition and subtraction of 1 and 10.

©2018 Great Minds®. eureka-math.org

EUREKA MATH

Name _____   Date _____

1. Complete each pattern.

   a. 48, 47, 46, 45, 44, __45__, __46__, __47__

   b. 78, 68, 58, 48, 38, __39__, __40__, __40__

   c. 35, 34, 44, 43, 53, __54__, __55__, __56__

2. Create two patterns using one of these rules for each: +1, -1, +10, or -10.

   a. __20__, __30__, __40__, __50__

   Rule for Pattern (a): __60 70__

   b. __80__, __90__, __100__, __101__

   Rule for Pattern (b): __8+6=__

EUREKA MATH™

Lesson 1:   Relate 1 more, 1 less, 10 more, and 10 less to addition and subtraction of 1 and 10.

©2018 Great Minds®. eureka-math.org

7

unlabeled tens place value chart

Lesson 1: Relate 1 more, 1 less, 10 more, and 10 less to addition and subtraction of 1 and 10.

©2018 Great Minds®. eureka-math.org

9

R (Read the problem carefully.)

Susan has 57 cents in her piggy bank. If she just put in 30 cents today, how much did she have yesterday?

D (Draw a picture.)

W (Write and solve an equation.)

Lesson 2:   Add and subtract multiples of 10 including counting on to subtract.

©2018 Great Minds®. eureka-math.org

11

W (Write a statement that matches the story.)

_____

_____

_____

Add and subtract multiples of 10 including counting on to subtract.

EUREKA MATH

Name _____   Date _____

1. Solve using place value strategies. Use your personal white board to show the arrow way or number bonds, or just use mental math, and record your answers.

   a. 5 tens + 3 tens = ___8___ tens   2 tens + 7 tens = ___9___ tens

         50 + 30 = __80__   20 + 70 = __90__

   b. 24 + 30 = __54__   50 + 24 = __74__   14 + 50 = __45__

   c. 20 + 37 = __57__   37 + 40 = __77__   60 + 27 = __67__

   d. 57 + ____ = 87   ____ + 34 = 74   19 + ____ = 69

   e. ____ + 56 = 86   38 + ____ = 78   12 + ____ = 72

2. Solve using place value strategies.

   a. 8 tens – 2 tens = __6__ tens   7 tens – 3 tens = _____ tens

         80 – 20 = __60__   70 – 30 = _____

   b. 78 - 40 = _____   56 – 30 = _____   88 – 50 = _____

   c. 84 - ____ = 24   57 - ____ = 37   93 - ____ = 43

   d. 83 - ____ = 23   54 - ____ = 34   91 - ____ = 41

3. Solve.

   a. 39 + _____ = 69

   b. 8 tens 7 ones – 3 tens = _____

   c. _____ + 5 tens = 7 tens

   d. _____ + 5 tens 6 ones = 8 tens 6 ones

   e. 48 ones – 2 tens = _____ tens _____ ones

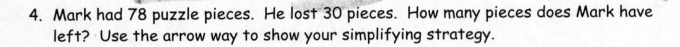

4. Mark had 78 puzzle pieces. He lost 30 pieces. How many pieces does Mark have left? Use the arrow way to show your simplifying strategy.

Lesson 2:   Add and subtract multiples of 10 including counting on to subtract.

EUREKA MATH™

Name _____ Date _____

Fill in the missing number to make each statement true.

1. 50 + 20 = _____

2. 4 tens + 3 tens = _____ tens

3. 7 tens - _____ tens = 5 tens

4. _____ - 20 = 63

5. 6 tens + 1 ten 4 ones = 9 tens 4 ones - _____ tens

EUREKA
MATH™

Lesson 2:   Add and subtract multiples of 10 including counting on to subtract.

©2018 Great Minds®. eureka-math.org

15

R (Read the problem carefully.)

Terrell put 19 stamps in his book on Monday. On Tuesday, he put in 32 stamps.

a. How many stamps did Terrell put in his book on Monday and Tuesday?

b. If Terrell's book holds 90 stamps, how many more stamps does he need to fill his book?

D (Draw a picture.)

W (Write and solve an equation.)

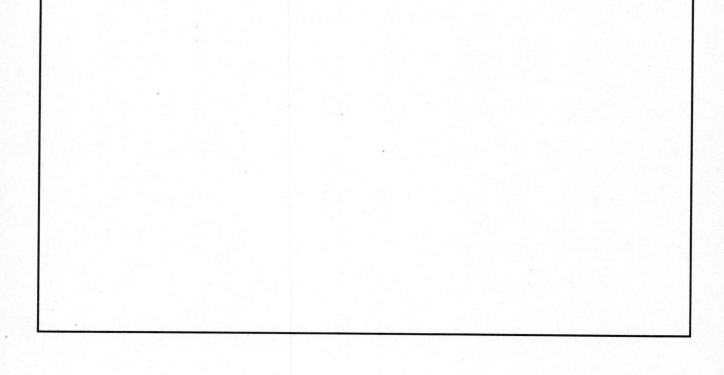

Lesson 3: Add and subtract multiples of 10 and some ones within 100.

©2018 Great Minds®. eureka-math.org

17

W (Write a statement that matches the story.)

a. _____

_____

b. _____

_____

Lesson 3: Add and subtract multiples of 10 and some ones within 100.

©2018 Great Minds®. eureka-math.org

EUREKA MATH

Name _____ Date _____

1. Solve each using the arrow way.

a.

38 + 20 58

38 + 21 59

38 + 19 57

b.

47 + 40 51

t answer

47 + 41 88

47 + 39 86

c.

34 – 10 24

34 – 11 23

34 – 9 25

d.

45 – 20 24

45 – 21 54

45 – 19 34

2. Solve using the arrow way, number bonds, or mental math. Use scrap paper if needed.

a. 49 + 20 = _____	21 + 49 = _____	49 + 19 = _____
b. 23 + 70 = _____	23 + 71 = _____	69 + 23 = _____
c. 84 – 20 = _____	84 – 21 = _____	84 – 19 = _____
d. 94 – 41 = _____	94 – 39 = _____	94 – 37 = _____
e. 73 – 29 = _____	52 – 29 = _____	85 – 29 = _____

3. Jessie's mom buys snacks for his classroom. She buys 22 apples, 19 oranges, and 49 strawberries. How many pieces of fruit does Jessie's mom buy?

Lesson 3:   Add and subtract multiples of 10 and some ones within 100.

EUREKA MATH

Name _____ Date _____

1. Solve using the arrow way or number bonds.

   a. 43 + 30 = _____

   b. 68 + 24 = _____

   c. 82 – 51 = _____

   d. 28 – 19 = _____

2. Show or explain how you used mental math to solve one of the problems above.

EUREKA MATH

Lesson 3:   Add and subtract multiples of 10 and some ones within 100.

©2018 Great Minds®. eureka-math.org

21

R (Read the problem carefully.)

Carlos bought 61 t-shirts. He gave 29 of them to his friends. How many t-shirts does Carlos have left?

D (Draw a picture.)

W (Write and solve an equation.)

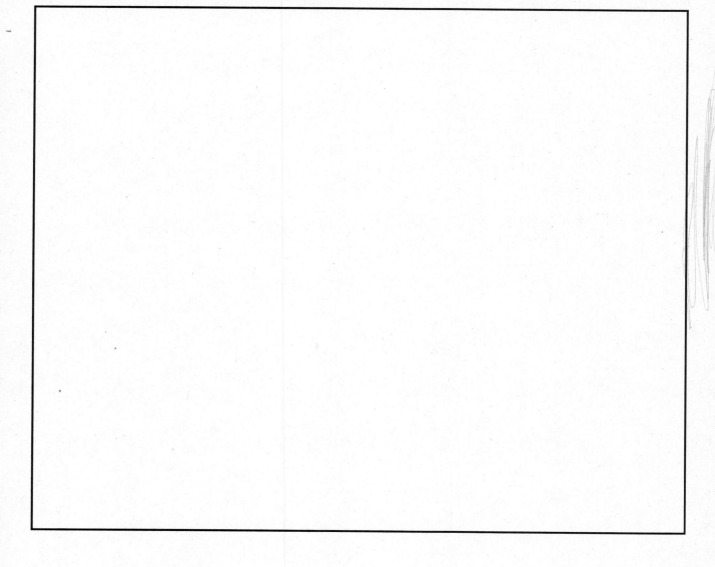

EUREKA MATH™

Lesson 4: Add and subtract multiples of 10 and some ones within 100.

©2018 Great Minds®. eureka-math.org

23

W (Write a statement that matches the story.)

_____

_____

_____

Lesson 4:   Add and subtract multiples of 10 and some ones within 100.

EUREKA MATH

Name _____   Date _____

1. Solve. Draw and label a tape diagram to subtract tens. Write the new number sentence.

   a. 23 – 9 = ___24 – 10___ = _____

   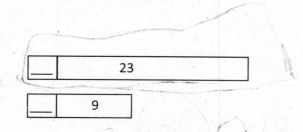

   b. 32 – 19 = _____ = _____

   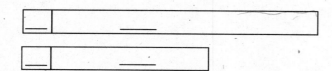

   c. 50 – 29 = _____ = _____

   d. 47 – 28 = _____ = _____

2. Solve. Draw and label a tape diagram to add tens. Write the new number sentence.

a. 29 + 46 = ___30 + 45___ = _____

29	1	45

b. 38 + 45 = _____ = _____

c. 61 + 29 = _____ = _____

d. 27 + 68 = _____ = _____

Lesson 4:   Add and subtract multiples of 10 and some ones within 100.

EUREKA MATH™

Name _____   Date _____

1. Solve. Draw a tape diagram or number bond to add or subtract tens. Write the new number sentence.

   a. 26 + 38 = _____ = _____

   b. 83 – 46 = _____ = _____

2. Craig checked out 28 books at the library. He read and returned some books. He still has 19 books checked out. How many books did Craig return? Draw a tape diagram or number bond to solve.

Name _____   Date _____

Solve and show your strategy.

1. 39 books were on the top bookshelf. Marcy added 48 more books to the top shelf. How many books are on the top shelf now?

$$39 + 48 = 87 \qquad 47 + 40 = 87$$

39 49 59 69 79

79

87

$$\begin{array}{r} 47 \\ + 40 \\ \hline 87 \end{array}$$

2. There are 53 regular pencils and some colored pencils in the bin. There are a total of 91 pencils in the bin. How many colored pencils are in the bin?

$$91 - 53 = \boxed{38}$$

7 + 30 + 1

91

53

$$7 + 1 = 8 \qquad 10 - 3 = 7$$

$$\begin{array}{r} 30 \\ + 8 \\ \hline \boxed{38} \end{array}$$

53 63 73 83 93

91

$$\begin{array}{r} 12 \\ 3 \cancel{8} \end{array}$$

Lesson 5:   Solve one- and two-step word problems within 100 using strategies based on place value.

29

©2018 Great Minds®. eureka-math.org

3. Henry solved 24 of his homework problems. There were 51 left to do. How many math problems were there on his homework sheet?

4. Matthew has 68 stickers. His brother has 29 fewer stickers.

   a. How many stickers does Matthew's brother have?

   b. How many stickers do Matthew and his brother have altogether?

Lesson 5:   Solve one- and two-step word problems within 100 using strategies based on place value.

EUREKA MATH

5. There are 47 photos in the blue album. The blue album has 32 more photos than the red album.

   a. How many photos are in the red album?

   b. How many photos are in the red and blue albums altogether?

6. Kiera has 62 blocks, and Pete has 37 blocks. They give away 75 blocks. How many blocks do they have left?

Lesson 5:   Solve one- and two-step word problems within 100 using strategies based on place value.

©2018 Great Minds®. eureka-math.org

31

Name _____ Date _____

Solve and show your strategy.

1. A store sold 58 t-shirts and had 25 t-shirts left.

   a. How many t-shirts did the store have at first?

   b. If 17 t-shirts are returned, how many t-shirts does the store have now?

2. Steve swam 23 laps in the pool on Saturday, 28 laps on Sunday, and 36 laps on Monday. How many laps did Steve swim?

Lesson 5:   Solve one- and two-step word problems within 100 using strategies
            based on place value.

©2018 Great Minds®. eureka-math.org

33

R (Read the problem carefully.)

Mr. Wally's class collects 36 cans for the recycling program. Then, Azniv brings in 8 more cans. How many cans does the class have now?

D (Draw a picture.)

W (Write and solve an equation.)

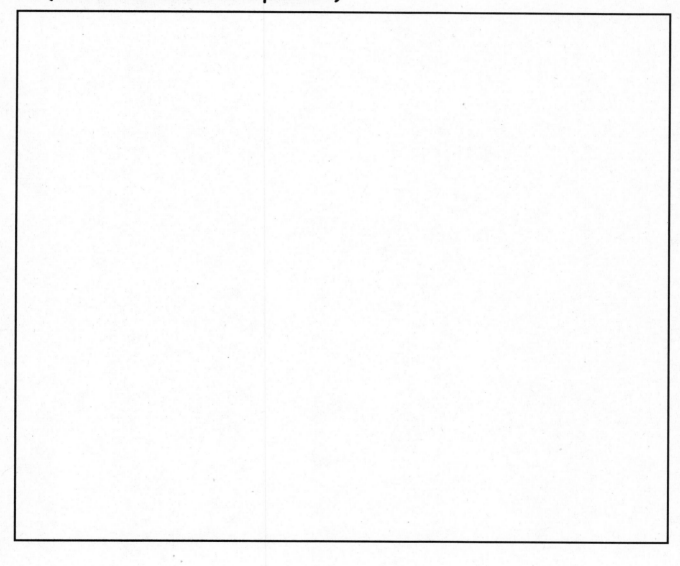

 EUREKA MATH™

Lesson 6:   Use manipulatives to represent the composition of 10 ones as 1 ten with two-digit addends.

©2018 Great Minds®. eureka-math.org

35

W (Write a statement that matches the story.)

_____

_____

_____

Lesson 6:   Use manipulatives to represent the composition of 10 ones as 1 ten
with two-digit addends.

EUREKA MATH™

Name _____   Date _____

1. Solve using mental math, if you can. Use your place value chart and place value disks to solve those you cannot solve mentally.

    a. 6 + 8 = __14__   30 + 8 = __38__   36 + 8 = __44__   36 + 48 = __84__

    b. 5 + 7 = __13__   20 + 7 = __27__   25 + 7 = __32__   25 + 57 = _____

                                                          70

2. Solve the following problems using your place value chart and place value disks. Compose a ten, if needed. Think about which ones you can solve mentally, too!

    a. 35 + 5 = __40__   35 + 6 = __41__
        30 + 10 =   40

    b. 26 + 4 = __30__   26 + 5 = __30__
        30 + 0   30

    c. 54 + 15 = __69__   54 + 18 = __70__
        50 + 20   50 + 20

    d. 67 + 23 = __90__   67 + 25 = __100__
        70 + 20   70 + 30

    e. 45 + 26 = __80__   45 + 23 = __70__
        50 + 30   50 + 20

    f. 58 + 23 = __80__   58 + 25 = __80__
        60 + 20   60 + 20

    g. 49 + 37 = __80__   52 + 36 = __80__
        50 + 40   50 + 30

3. There are 47 blue buttons and 25 black buttons in Sean's drawer. How many buttons are in his drawer?

For early finishers:

4. Leslie has 24 blue and 24 pink hair ribbons. She buys 17 more blue ribbons and 13 more pink ribbons from the store.

   a. How many blue hair ribbons does she have now?

   b. How many pink hair ribbons does she have now?

   c. Jada has 29 more pink ribbons than Leslie. How many pink ribbons does Jada have?

Lesson 6:   Use manipulatives to represent the composition of 10 ones as 1 ten
                                with two-digit addends.

EUREKA
MATH™

Name _____   Date _____

Solve using your place value chart and place value disks. Compose a ten, if needed.
Think about which ones you can solve mentally, too!

1. 53 + 19 = _____

2. 44 + 27 = _____

3. 64 + 28 = _____

R (Read the problem carefully.)

Farmer Andino's chickens laid 47 brown eggs and 39 white eggs. How many eggs did the chickens lay in all?

D (Draw a picture.)

W (Write and solve an equation.)

W (Write a statement that matches the story.)

_____

_____

_____

Lesson 7:   Relate addition using manipulatives to a written vertical method.

©2018 Great Minds®. eureka-math.org

EUREKA MATH

Name _____   Date _____

1. Solve the following problems using the vertical form, your place value chart, and place value disks. Bundle a ten, when necessary. Think about which ones you can solve mentally, too!

   a. $22 + 8 = 30$ $21 + 9$ 30

   b. $34 + 17$ 51 $33 + 18 = 51$

   c. $48 + 34$ 82 $46 + 36$ 80

   d. $27 + 68$ 95 $26 + 69$ 95

Extra Practice for Early Finishers: Solve the following problems using your place value chart and place value disks. Bundle a ten, when necessary.

2. Samantha brought grapes to school for a snack. She had 27 green grapes and 58 red grapes. How many grapes did she bring to school?

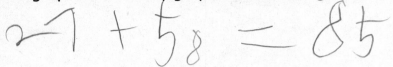

$$27 + 58 = 85$$

3. Thomas read 29 pages of his new book on Monday. On Tuesday, he read 35 more pages than he did on Monday.

   a. How many pages did Thomas read on Tuesday?

   b. How many pages did Thomas read on both days?

EUREKA MATH

Name _____   Date _____

1. Solve the following problems using the vertical form, your place value chart, and place value disks. Bundle a ten, if needed. Think about which ones you can solve mentally, too!

   a. 47 + 34

   b. 54 + 27

2. Explain how Problem 1, Part (a) can help you solve Problem 1, Part (b).

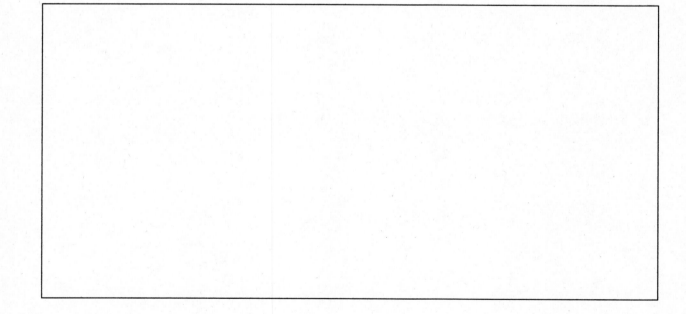

R (Read the problem carefully.)

At the school fair, 29 cupcakes were sold, and 19 were left over. How many cupcakes were brought to the fair?

D (Draw a picture.)

W (Write and solve an equation.)

Lesson 8:   Use math drawings to represent the composition and relate drawings
to a written method.

©2018 Great Minds®. eureka-math.org

47

W (Write a statement that matches the story.)

_____

_____

_____

Lesson 8:   Use math drawings to represent the composition and relate drawings to a written method.

©2018 Great Minds®. eureka-math.org

EUREKA MATH™

Name _____ Date _____

1. Solve vertically. Draw and bundle place value disks on the place value chart.

   a. 27 + 15 = _____

   b. 44 + 26 = _____

   c. 48 + 31 = _____

   d. 33 + 59 = _____

EUREKA MATH

Lesson 8: Use math drawings to represent the composition and relate drawings to a written method.

©2018 Great Minds®. eureka-math.org

49

e. 27 + 45 = _____

f. 18 + 68 = _____

2. There are 23 laptops in the computer room and 27 laptops in the first-grade classroom. How many laptops are in the computer room and first-grade classroom altogether?

For early finishers:

3. Mrs. Anderson gave 36 pencils to her class and had 48 left over. How many pencils did Mrs. Anderson have at first?

Lesson 8: Use math drawings to represent the composition and relate drawings to a written method.

©2018 Great Minds®. eureka-math.org

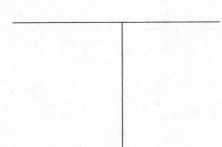

EUREKA MATH™

Name _____   Date _____

Use place value language to explain Zane's mistake. Then, solve using the vertical form. Draw and bundle place value disks on your place value chart.

Zane's Answer	Zane's Mistake
59 + 35 = _____	

My Answer

EUREKA MATH™

Lesson 8: Use math drawings to represent the composition and relate drawings to a written method.

©2018 Great Minds®. eureka-math.org

51

R (Read the problem carefully.)

Maria spilled a box of paper clips. They landed on her desk and on the floor. 20 of them landed on her desk. Five more fell on the floor than landed on her desk. How many paper clips did she spill?

D (Draw a picture.)

W (Write and solve an equation.)

Lesson 9:   Use math drawings to represent the composition when adding a two-digit to a three-digit addend.

©2018 Great Minds®. eureka-math.org

53

W (Write a statement that matches the story.)

_____

_____

_____

Lesson 9:   Use math drawings to represent the composition when adding a two-digit to a three-digit addend.

©2018 Great Minds®. eureka-math.org

EUREKA MATH™

Name _____   Date _____

1. Solve using the algorithm. Draw and bundle chips on the place value chart.

a. 123 + 16 = _____

$$\begin{array}{r} 123 \\ +\ 16 \\ \hline 139 \end{array}$$

hundreds	tens	ones

b. 111 + 79 = _____

hundreds	tens	ones

c. 109 + 33 = _____

hundreds	tens	ones

EUREKA MATH

Lesson 9: Use math drawings to represent the composition when adding a two-digit to a three-digit addend.

©2018 Great Minds®. eureka-math.org

55

d. 57 + 138 = _____

hundreds	tens	ones

2. Jose sold 127 books in the morning. He sold another 35 books in the afternoon. At the end of the day he had 19 books left.

a. How many books did Jose sell?

hundreds	tens	ones

b. How many books did Jose have at the beginning of the day?

hundreds	tens	ones

Lesson 9: Use math drawings to represent the composition when adding a two-digit to a three-digit addend.

EUREKA MATH

Name _____ Date _____

1. Solve using the algorithm. Write a number sentence for the problem modeled on the place value chart.

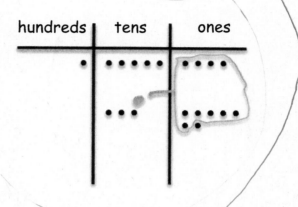

2. Solve using the algorithm. Draw and bundle chips on the place value chart.

   136 + 39 = _____

hundreds	tens	ones

EUREKA MATH

Lesson 9: Use math drawings to represent the composition when adding a two-digit to a three-digit addend.

©2018 Great Minds®. eureka-math.org

57

R (Read the problem carefully.)

Moises sold 24 raffle tickets on Monday and 4 fewer tickets on Tuesday.

How many tickets did he sell in all on both days?

D (Draw a picture.)

W (Write and solve an equation.)

Lesson 10:   Use math drawings to represent the composition when adding a
two-digit to a three-digit addend.

©2018 Great Minds®. eureka-math.org

59

W (Write a statement that matches the story.)

_____

_____

_____

Lesson 10: Use math drawings to represent the composition when adding a two-digit to a three-digit addend.

EUREKA MATH

Name _____   Date _____

1. Solve using the algorithm. Draw chips and bundle when you can.

a. 127 + 18 = _____

hundreds	tens	ones

b. 136 + 16 = _____

hundreds	tens	ones

c. 109 + 41 = _____

hundreds	tens	ones

d. 29 + 148 = _____

hundreds	tens	ones

EUREKA MATH™

Lesson 10:   Use math drawings to represent the composition when adding a
two-digit to a three-digit addend.

©2018 Great Minds®. eureka-math.org

61

e. 79 + 107 = _____

hundreds	tens	ones

Before bundling a ten   _____ hundreds   _____ tens   _____ ones

After bundling a ten   _____ hundreds   _____ tens   _____ ones

2. a. On Saturday, Colleen earned 4 ten-dollar bills and 18 one-dollar bills working on the farm. How much money did Colleen earn?

hundreds	tens	ones

b. On Sunday, Colleen earned 2 ten-dollar bills and 16 one-dollar bills. How much money did she earn on both days?

hundreds	tens	ones

Lesson 10:   Use math drawings to represent the composition when adding a
two-digit to a three-digit addend.

©2018 Great Minds®. eureka-math.org

EUREKA
MATH

Name _____ Date _____

1. Solve using the algorithm. Draw chips and bundle when you can.

    27 + 137

hundreds	tens	ones

2. Using the previous problem, fill in the blanks. Use place value language to explain how you used bundling to rename the solution.

    Before bundling a ten   _____ hundreds   _____ tens   _____ ones

    After bundling a ten   _____ hundreds   _____ tens   _____ ones

    <u>Explanation</u>

EUREKA MATH

Lesson 10:   Use math drawings to represent the composition when adding a two-digit to a three-digit addend.

©2018 Great Minds®. eureka-math.org

63

R (Read the problem carefully.)

Shelby picks 35 oranges. 5 are rotten.

   a. How many of Shelby's oranges are not rotten?

   b. Rosa picks 35 oranges, too, but 6 are rotten. How many of Rosa's

   oranges are not rotten?

D (Draw a picture.)

W (Write and solve an equation.)

Lesson 11:   Represent subtraction with and without the decomposition of
1 ten as 10 ones with manipulatives.

©2018 Great Minds®. eureka-math.org

65

W (Write a statement that matches the story.)

a. _____

_____

b. _____

_____

Lesson 11:   Represent subtraction with and without the decomposition of
1 ten as 10 ones with manipulatives.

©2018 Great Minds®. eureka-math.org

EUREKA
MATH™

Name _____   Date _____

1. Solve using mental math.

    a. 8 – 7 = _____   38 – 7 = _____   38 – 8 = _____   38 – 9 = _____

    b. 7 – 6 = _____   87 – 6 = _____   87 – 7 = _____   87 – 8 = _____

2. Solve using your place value chart and place value disks. Unbundle a ten if needed.
    Think about which problems you can solve mentally, too!

    a. 28 – 7 = _____   28 – 9 = _____

    b. 25 – 5 = _____   25 – 6 = _____

    c. 30 – 5 = _____   33 – 5 = _____

    d. 47 – 22 = _____   41 – 22 = _____

    e. 44 – 16 = _____   44 – 26 = _____

    f. 70 – 28 = _____   80 – 28 = _____

EUREKA
MATH

Lesson 11:   Represent subtraction with and without the decomposition of
              1 ten as 10 ones with manipulatives.

©2018 Great Minds®. eureka-math.org

67

3. Solve 56 – 28, and explain your strategy.

For early finishers:

4. There are 63 problems on the math test. Tamara answered 48 problems correctly, but the rest were incorrect. How many problems did she answer incorrectly?

5. Mr. Ross has 7 fewer students than Mrs. Jordan. Mr. Ross has 35 students. How many students does Mrs. Jordan have?

Lesson 11:   Represent subtraction with and without the decomposition of
1 ten as 10 ones with manipulatives.

©2018 Great Minds®. eureka-math.org

EUREKA MATH

Name _____   Date _____

Solve for the missing part. Use your place value chart and place value disks.

1.

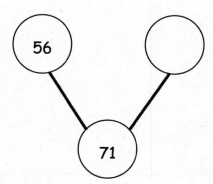

2.

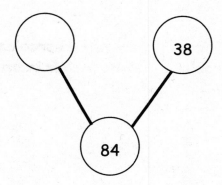

EUREKA MATH™   **Lesson 11:**   Represent subtraction with and without the decomposition of   69
                                      1 ten as 10 ones with manipulatives.

©2018 Great Minds®. eureka-math.org

R (Read the problem carefully.)

Barb has a bag of 34 cherries. She eats 17 cherries for a snack. How many cherries does she have left?

D (Draw a picture.)

W (Write and solve an equation.)

W (Write a statement that matches the story.)

_____

_____

_____

EUREKA MATH™

Name _____   Date _____

1. Use place value disks to solve each problem. Rewrite the problem vertically, and record each step as shown in the example.

   a. 22 – 18

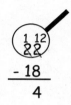

   b. 20 – 12

   c. 34 – 25

   d. 25 – 18

   e. 53 – 29

   f. 71 – 27

2. Terry and Pam both solved the problem 64 – 49. They came up with different answers and cannot agree on who is correct. Terry answered 25, and Pam answered 15. Use place value disks to explain who is correct, and rewrite the problem vertically to solve.

For early finishers:

3. Samantha has 42 marbles, and Graham has 17 marbles.

   a. How many more marbles does Samantha have than Graham?

   b. James has 25 fewer marbles than Samantha. How many marbles does James have?

Lesson 12:   Relate manipulative representations to a written method.

EUREKA
MATH™

Name _____   Date _____

Sherry made a mistake while subtracting. Explain her mistake.

Sherry's Work:	Explanation:
14 4̶4̶ −26 28	_____ _____ _____ _____ _____ _____

R (Read the problem carefully.)

Mrs. Beachy went shopping with $42. She spent $18. How much money did she have left?

D (Draw a picture.)

W (Write and solve an equation.)

Lesson 13:   Use math drawings to represent subtraction with and without
              decomposition and relate drawings to a written method.

©2018 Great Minds®. eureka-math.org

77

W (Write a statement that matches the story.)

_____

_____

_____

Lesson 13:   Use math drawings to represent subtraction with and without decomposition and relate drawings to a written method.

EUREKA MATH

Name _____   Date _____

1. Solve vertically. Draw a place value chart and chips to model each problem.
    Show how you change 1 ten for 10 ones, when necessary.

a. 31 – 19 = _____	b. 46 – 24 = _____
c. 51 – 33 = _____ 	d. 67 – 49 = _____
e. 66 – 48 = _____ 5 8̸16 – 48 ——— 18	f. 77 – 58 = _____ 6 7̸7̸ – 58 ———

EUREKA MATH

Lesson 13:   Use math drawings to represent subtraction with and without
decomposition and relate drawings to a written method.

©2018 Great Minds®. eureka-math.org

79

2. Solve 31 – 27 and 25 – 15 vertically using the space below. Circle to tell if the number sentence is true or false.

True or False

31 – 27   =   25 – 15

3. Solve 78 – 43 and 81 – 46 vertically using the space below. Circle to tell if the number sentence is true or false.

True or False

78 – 43   =   81 – 46

4. Mrs. Smith has 39 tomatoes in her garden. Mrs. Thompson has 52 tomatoes in her garden. How many fewer tomatoes does Mrs. Smith have than Mrs. Thompson?

Lesson 13:   Use math drawings to represent subtraction with and without decomposition and relate drawings to a written method.

©2018 Great Minds®. eureka-math.org

EUREKA MATH™

Name _____ Date _____

Solve vertically. Draw a place value chart and chips to model each problem. Show how you change 1 ten for 10 ones, when necessary.

1. 75 – 28 = _____

2. 63 – 35 = _____

Lesson 13: Use math drawings to represent subtraction with and without decomposition and relate drawings to a written method.

©2018 Great Minds®. eureka-math.org

81

R (Read the problem carefully.)

The total length of a red string and a purple string is 73 cm. The red string is 18 cm long. How long is the purple string?

Extension: Find the difference in length between the two strings.

D (Draw a picture.)

W (Write and solve an equation.)

Lesson 14:   Represent subtraction with and without the decomposition when there is a three-digit minuend.

©2018 Great Minds®. eureka-math.org

W (Write a statement that matches the story.)

_____

_____

_____

Lesson 14: Represent subtraction with and without the decomposition when there is a three-digit minuend.

EUREKA MATH

Name _____   Date _____

1. Solve by writing the problem vertically. Check your result by drawing chips on the place value chart. Change 1 ten for 10 ones, when needed.

    a. 134 – 23= _____

hundreds	tens	ones

    b. 140 – 12 = _____

hundreds	tens	ones

    c. 121 – 14 = _____

hundreds	tens	ones

EUREKA MATH™   Lesson 14:   Represent subtraction with and without the decomposition when there is a three-digit minuend.   85

©2018 Great Minds®. eureka-math.org

d. 161 – 26 = _____

hundreds	tens	ones

e. 187 – 49 = _____

hundreds	tens	ones

2. Solve the following problems vertically without a place value chart.

a. 63 – 28 = _____	b. 163 – 28 = _____

Lesson 14: Represent subtraction with and without the decomposition when there is a three-digit minuend.

EUREKA MATH

Name _____   Date _____

Solve by writing the problem vertically. Check your result by drawing chips on the place value chart. Change 1 ten for 10 ones, when needed.

1. 145 – 28 = _____

$$
\begin{array}{r}
1\ 4\ 1\ 5 \\
-\ 2\ 8 \\
\hline
1\ 3\ 5
\end{array}
$$

hundreds	tens	ones

2. 151 – 39 = _____

hundreds	tens	ones

EUREKA MATH™   Lesson 14:   Represent subtraction with and without the decomposition when there is a three-digit minuend.   87

©2018 Great Minds®. eureka-math.org

R (Read the problem carefully.)

There are 136 students in the second grade at Miles Davis Elementary.

27 of them brought bag lunches to school. The rest buy the hot lunch.

How many students are buying a hot lunch?

D (Draw a picture.)

W (Write and solve an equation.)

Lesson 15: Represent subtraction with and without the decomposition when there is a three-digit minuend.

©2018 Great Minds®. eureka-math.org

89

W (Write a statement that matches the story.)

_____

_____

_____

       Lesson 15:   Represent subtraction with and without the decomposition when there is a three-digit minuend.

Name _____   Date _____

1. Solve each problem using vertical form. Show the subtraction on the place value chart with chips. Exchange 1 ten for 10 ones, when necessary.

a. 173 – 42

hundreds	tens	ones

b. 173 – 38

hundreds	tens	ones

c. 170 – 44

hundreds	tens	ones

EUREKA
MATH™

Lesson 15: Represent subtraction with and without the decomposition when there is a three-digit minuend.

91

©2018 Great Minds®. eureka-math.org

d. 150 – 19

hundreds	tens	ones
150		
79	1 3	2

e. 186 – 57

hundreds	tens	ones

2. Solve the following problems without using a place value chart.

a. 73 – 56	b. 170 – 53

Lesson 15:   Represent subtraction with and without the decomposition when
there is a three-digit minuend.

©2018 Great Minds®. eureka-math.org

EUREKA
MATH™

Name _____   Date _____

Solve using vertical form. Show the subtraction on a place value chart with chips. Exchange 1 ten for 10 ones, when necessary.

1. 164 – 49

hundreds	tens	ones

2. 181 – 73

hundreds	tens	ones

EUREKA MATH™

Lesson 15:   Represent subtraction with and without the decomposition when there is a three-digit minuend.

©2018 Great Minds®. eureka-math.org

93

Name _____ Date _____

Solve the following word problems. Use the RDW process.

1. Frederick counted a total of 80 flowers in the garden. There were 39 white flowers, and the rest were pink. How many flowers were pink?

2. The clothing store had 42 shirts. After selling some, there were 16 left. How many shirts were sold?

3. There were 26 magazines on Shelf A and 60 magazines on Shelf B. How many more magazines were on Shelf B than Shelf A?

EUREKA MATH

Lesson 16:   Solve one- and two-step word problems within 100 using strategies based on place value.

©2018 Great Minds®. eureka-math.org

95

4. Andy spent 71 hours studying in November.

   In December, he studied 19 hours less.
   Rachel studied 22 hours more than Andy studied in December.
   How many hours did Rachel study in December?

5. Thirty-six books are in the blue bin.

   The blue bin has 18 more books than the red bin.
   The yellow bin has 7 more books than the red bin.

   a. How many books are in the red bin?

   b. How many books are in the yellow bin?

Lesson 16: Solve one- and two-step word problems within 100 using strategies based on place value.

©2018 Great Minds®. eureka-math.org

EUREKA MATH

Name _____   Date _____

Solve the following word problems. Use the RDW process.

1. The bookstore sold 83 books on Monday.
    On Tuesday, it sold 46 fewer books than on Monday.

    a. How many books were sold on Tuesday?

    b. The bookstore sold 28 more books on Tuesday than on Wednesday.
        How many books did the bookstore sell on Wednesday?

Lesson 16:   Solve one- and two-step word problems within 100 using strategies
based on place value.

©2018 Great Minds®. eureka-math.org

97

R (Read the problem carefully.)

Erasers come in boxes of 10. Victor has 14 boxes. Gabby has 5 boxes.

a. How many erasers does Victor have?

b. How many erasers does Gabby have?

c. If Gabby gets another box, how many erasers do they have in all?

D (Draw a picture.)

W (Write and solve an equation.)

EUREKA MATH

Lesson 17: Use mental strategies to relate compositions of 10 tens as 1 hundred to 10 ones as 1 ten.

©2018 Great Minds®. eureka-math.org

99

W (Write a statement that matches the story.)

a. _____

_____

b. _____

_____

c. _____

_____

Lesson 17: Use mental strategies to relate compositions of 10 tens as 1 hundred to 10 ones as 1 ten.

©2018 Great Minds®. eureka-math.org

EUREKA MATH

Name _____   Date _____

1. Solve mentally.

   a. 2 ones + _____ = 1 ten   2 + _____ = 10

      2 tens + _____ = 1 hundred   20 + _____ = 100

   b. 1 ten = _____ + 6 ones   10 = _____ + 6

      1 hundred = _____ + 6 tens   100 = _____ + 60

   c. 3 ones + 7 ones = _____ ten   3 + 7 = _____

      3 tens + 7 tens = _____ tens   30 + 70 = _____

      13 tens + 7 tens = _____ tens   130 + 70 = _____

   d. 6 ones + 4 ones = _____ ten   6 + 4 = _____

      16 tens + 4 tens = _____ hundreds   160 + 40 = _____

   e. 12 ones + 8 ones = _____ tens   12 + 8 = _____

      12 tens + 8 tens = _____ hundreds   120 + 80 = _____

2. Solve.

a. 9 ones + 4 ones = _____ ten _____ ones $9 + 4 =$ _____

9 tens + 4 tens = _____ hundred _____ tens $90 + 40 =$ _____

b. 4 ones + 8 ones = _____ ten _____ ones $4 + 8 =$ _____

4 tens + 8 tens = _____ hundred _____ tens $40 + 80 =$ _____

c. 6 ones + 7 ones = _____ ten _____ ones $6 + 7 =$ _____

6 tens + 7 tens = _____ hundred _____ tens $60 + 70 =$ _____

3. Fill in the blanks. Then, complete the addition sentence.
The first one is done for you.

a. $24 \xrightarrow{+6}$ __30__ $\xrightarrow{+70}$ __100__

24 + __76__ = __100__

b. $124 \xrightarrow{+6}$ _____ $\xrightarrow{+70}$ _____

124 + _____ = _____

c. $7 \xrightarrow{+3}$ _____ $\xrightarrow{+90}$ _____ $\xrightarrow{+100}$ _____

7 + _____ = _____

d. $70 \xrightarrow{+30}$ _____ $\xrightarrow{+90}$ _____ $\xrightarrow{+10}$ _____

70 + _____ = _____

e. $38 \xrightarrow{+2}$ _____ $\xrightarrow{+60}$ _____ $\xrightarrow{+30}$ _____

38 + _____ = _____

f. $98 \xrightarrow{+2}$ _____ $\xrightarrow{+6}$ _____ $\xrightarrow{+40}$ _____

98 + _____ = _____

Lesson 17:   Use mental strategies to relate compositions of 10 tens as
1 hundred to 10 ones as 1 ten.

©2018 Great Minds®. eureka-math.org

EUREKA
MATH

Name _____ Date _____

1. Solve mentally.

   a. 4 ones + _____ = 1 ten   4 + _____ = 10

      4 tens + _____ = 1 hundred   40 + _____ = 100

   b. 2 ones + 8 ones = _____ ten   2 + 8 = _____

      2 tens + 18 tens = _____ hundreds   20 + 180 = _____

2. Fill in the blanks. Then, complete the addition sentence.

   63 $\xrightarrow{+7}$ _____ $\xrightarrow{+10}$ _____ $\xrightarrow{+10}$ _____ $\xrightarrow{+10}$ _____

   63 + _____ = _____

EUREKA MATH

Lesson 17:   Use mental strategies to relate compositions of 10 tens as
              1 hundred to 10 ones as 1 ten.

©2018 Great Minds®. eureka-math.org

103

Hailey and Gio solve 56 + 85. Gio says the answer is 131. Hailey says the answer is 141. Explain whose answer is correct using numbers, pictures, or words.

EUREKA MATH

Lesson 18:   Use manipulatives to represent additions with two compositions.

©2018 Great Minds®. eureka-math.org

105

Name _____ Date _____

1. Solve using your place value chart and place value disks.

   a. 80 + 30 = _110_   90 + 40 = _141_

   b. 73 + 38 = _111_   73 + 49 = _213_

   c. 93 + 38 = _131_   42 + 99 = _111_

   d. 84 + 37 = _123_   69 + 63 = _132_

   e. 113 + 78 = _193_   128 + 72 = _101_

2. Circle the statements that are true as you solve each problem using place value disks.

a. 47 + 123	b. 97 + 54
I change 10 ones for 1 ten.	I change 10 ones for 1 ten.
I change 10 tens for 1 hundred.	I change 10 tens for 1 hundred.
The total of the two parts is 160.	The total of the two parts is 141.
The total of the two parts is 170.	The total of the two parts is 151.

EUREKA MATH

Lesson 18:   Use manipulatives to represent additions with two compositions.

©2018 Great Minds®. eureka-math.org

107

3. Write an addition sentence that corresponds to the following number bond. Solve the problem using your place value disks, and fill in the missing total.

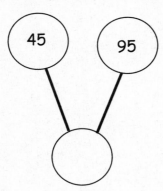

4. There are 50 girls and 80 boys in the after school program. How many children are in the after school program?

5. Kim and Stacy solved 83 + 39. Kim's answer was less than 120. Stacy's answer was more than 120. Whose answer was incorrect? Explain how you know using words, pictures, or numbers.

Lesson 18: Use manipulatives to represent additions with two compositions.

©2018 Great Minds®. eureka-math.org

EUREKA
MATH

Name _____   Date _____

Solve using your place value chart and place value disks.

1. 46 + 54 = _____

2. 49 + 56 = _____

3. 28 + 63 = _____

4. 67 + 89 = _____

EUREKA MATH

Lesson 18:   Use manipulatives to represent additions with two compositions.

109

©2018 Great Minds®. eureka-math.org

unlabeled hundreds place value chart

Lesson 18: Use manipulatives to represent additions with two compositions.

111

©2018 Great Minds®. eureka-math.org

R (Read the problem carefully.)

There are 35 note cards in one box. There are 67 note cards in another box. How many note cards are there in all?

D (Draw a picture.)

W (Write and solve an equation.)

W (Write a statement that matches the story.)

_____

_____

_____

Name _____   Date _____

1. Solve the following problems using the vertical form, your place value chart, and place value disks. Bundle a ten or hundred, if needed.

a. 72 + 19	b. 28 + 91
c. 68 + 61	d. 97 + 35
e. 68 + 75	f. 96 + 47

g. 177 + 23	h. 146 + 54

2. Thirty-eight fewer girls attended summer camp than boys. Seventy-nine girls attended.

   a. How many boys attended summer camp?

   b. How many children attended summer camp?

Lesson 19:   Relate manipulative representations to a written method.

EUREKA MATH

Name _____   Date _____

Solve the following problems using the vertical form, your place value chart, and place value disks. Bundle a ten or hundred, if needed.

1.   47 + 85

2.   128 + 39

Kendra and Jojo are counting their marbles. Kendra has 38, and Jojo has 62. Kendra says they have 100 marbles altogether, but Jojo says they have 90. Use words, numbers, or a model to prove who is correct.

Lesson 20:   Use math drawings to represent additions with up to two
compositions and relate drawings to a written method.

©2018 Great Minds®. eureka-math.org

119

Name _____   Date _____

1. Solve vertically. Draw chips on the place value chart and bundle, when needed.

a.   23 + 57 = _____

100's	10's	1's

b.   65 + 36 = _____

100's	10's	1's

c.   83 + 29 = _____

100's	10's	1's

EUREKA
MATH™

Lesson 20:   Use math drawings to represent additions with up to two
             compositions and relate drawings to a written method.

©2018 Great Minds®. eureka-math.org

121

d. 47 + 75 = _____

100's	10's	1's

e. 68 + 88 = _____

100's	10's	1's

2. Jessica's teacher marked her work incorrect for the following problem. Jessica cannot figure out what she did wrong. If you were Jessica's teacher, how would you explain her mistake?

Jessica's work:	Explanation:
100's 10's 1's $\begin{array}{r} 77 \\ + 32 \\ \hline 19 \end{array}$ 1 0 9	

Use math drawings to represent additions with up to two compositions and relate drawings to a written method.

©2018 Great Minds®. eureka-math.org

EUREKA MATH

Name _____ Date _____

Solve vertically. Draw chips on the place value chart and bundle, when needed.

1. 46 + 65 = _____

100's	10's	1's

2. 74 + 57 = _____

100's	10's	1's

EUREKA MATH

Lesson 20: Use math drawings to represent additions with up to two compositions and relate drawings to a written method.

©2018 Great Minds®. eureka-math.org

123

R (Read the problem carefully.)

Katrina has 23 stickers, and Jennifer has 9. How many more stickers does Jennifer need to have as many as Katrina?

D (Draw a picture.)

W (Write and solve an equation.)

Lesson 21:   Use math drawings to represent additions with up to two
compositions and relate drawings to a written method.

©2018 Great Minds®. eureka-math.org

125

W (Write a statement that matches the story.)

_____

_____

_____

Lesson 21: Use math drawings to represent additions with up to two compositions and relate drawings to a written method.

©2018 Great Minds®. eureka-math.org

EUREKA
MATH™

Name _____   Date _____

1. Solve vertically. Draw chips on the place value chart and bundle, when needed.

    a.   65 + 75 = __157__

100's	10's	1's

    b.   84 + 29 = __149__

100's	10's	1's

    c.   91 + 19 = __119__

100's	10's	1's

EUREKA
MATH™

©2018 Great Minds®. eureka-math.org

d. 163 + 27 = _____

100's	10's	1's

2. Abby solved 99 + 99 on her place value chart and in vertical form, but she got an incorrect answer. Check Abby's work, and correct it.

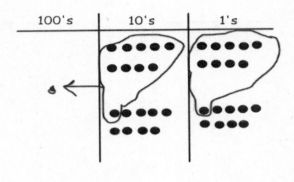

What did Abby do correctly?

_____

_____

What did Abby do incorrectly?

_____

_____

Lesson 21:   Use math drawings to represent additions with up to two
compositions and relate drawings to a written method.

©2018 Great Minds®. eureka-math.org

EUREKA
MATH™

Name _____   Date _____

Solve vertically. Draw chips on the place value chart and bundle, when needed.

1.   58 + 67 = _____

100's	10's	1's

2.   43 + 89 = _____

100's	10's	1's

EUREKA MATH

Lesson 21:   Use math drawings to represent additions with up to two
compositions and relate drawings to a written method.

©2018 Great Minds®. eureka-math.org

129

R (Read the problem carefully.)

There are 38 apples, 16 bananas, 24 peaches, and 12 pears in the fruit basket. How many pieces of fruit are in the basket?

D (Draw a picture.)

W (Write and solve an equation.)

Lesson 22: Solve additions with up to four addends with totals within 200 with
and without two compositions of larger units.

©2018 Great Minds®. eureka-math.org

131

W (Write a statement that matches the story.)

_____

_____

_____

Lesson 22:   Solve additions with up to four addends with totals within 200 with and without two compositions of larger units.

EUREKA MATH™

Name _____ Date _____

1. Look to make 10 ones or 10 tens to solve the following problems using place value strategies.

a. 5 + 5 + 7 = 17	25 + 25 + 17 = 47	125 + 25 + 17 = _____
b. 4 + 6 + 5 = 15	24 + 36 + 75 = _____	24 + 36 + 85 = _____
c. 2 + 4 + 8 + 6 = _____	32 + 24 + 18 + 46 = _____	72 + 54 + 18 + 26 = _____

EUREKA MATH

Lesson 22: Solve additions with up to four addends with totals within 200 with and without two compositions of larger units.

©2018 Great Minds®. eureka-math.org

133

2. Josh and Keith have the same problem for homework: 23 + 35 + 47 + 56. The students solved the problem differently but got the same answer.

Josh's work   Keith's work

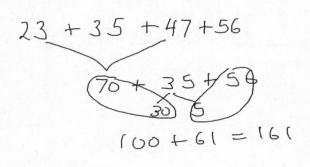

$100 + 61 = 161$

$60 + 101 = 161$

Solve 23 + 35 + 47 + 56 another way.

3. Melissa bought a dress for $29, a purse for $15, a book for $11, and a hat for $25. How much did Melissa spend? Show your work.

Lesson 22:   Solve additions with up to four addends with totals within 200 with and without two compositions of larger units.

©2018 Great Minds®. eureka-math.org

EUREKA MATH

Name _____ Date _____

Look to make 10 ones or 10 tens to solve the following problems using place value strategies.

1. 17 + 33 + 48

2. 35 + 56 + 89 + 18

EUREKA MATH

Lesson 22: Solve additions with up to four addends with totals within 200 with and without two compositions of larger units.

©2018 Great Minds®. eureka-math.org

135

R (Read the problem carefully.)

Yossef downloaded 115 songs. 100 of them were rock songs. The rest were hip-hop songs.

   a. How many of Yossef's songs were hip-hop?

   b. 80 of his rock songs were oldies rock. How many rock songs were new?

D (Draw a picture.)

W (Write and solve an equation.)

Lesson 23:   Use number bonds to break apart three-digit minuends and subtract from the hundred.

©2018 Great Minds®. eureka-math.org

137

W (Write a statement that matches the story.)

a. _____

_____

b. _____

_____

Lesson 23:   Use number bonds to break apart three-digit minuends and subtract
                                  from the hundred.

Name _____   Date _____

1. Solve using number bonds to subtract from 100. The first one has been done for you.

a. 106 – 90 = 16
 	
    6   100
 	
    100 – 90 = 10
    10 + 6 =   16
	b. 116 – 90

c. 114 – 80	d. 115 – 80
e. 123 – 70	f. 127 – 60

Lesson 23:   Use number bonds to break apart three-digit minuends and subtract from a hundred.

©2018 Great Minds

139

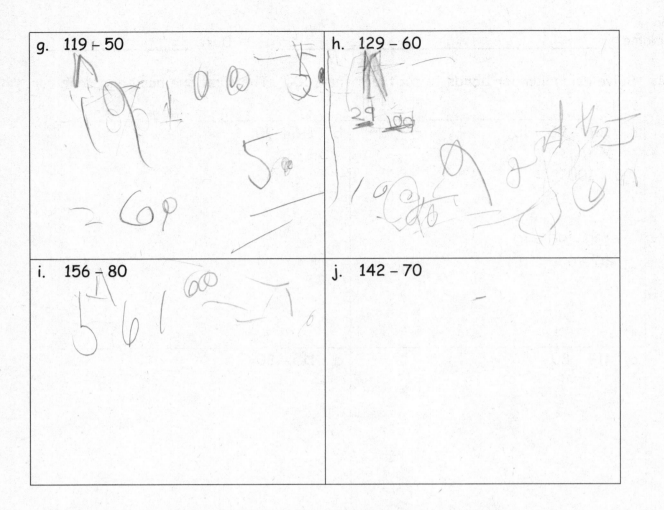

g. 119 – 50

h. 129 – 60

i. 156 – 80

j. 142 – 70

2. Use a number bond to show how you would take 8 tens from 126.

Lesson 23:   Use number bonds to break apart three-digit minuends and subtract
from the hundred.

EUREKA
MATH™

Name _____ Date _____

Solve using number bonds to subtract from 100.

1. 114 – 50

2. 176 – 90

3. 134 – 40

EUREKA MATH

Lesson 23: Use number bonds to break apart three-digit minuends and subtract from the hundred.

©2018 Great Minds®. eureka-math.org

141

R (Read the problem carefully.)

Sammy bought 114 notecards. He used 70 of them. How many unused notecards did he have left?

D (Draw a picture.)

W (Write and solve an equation.)

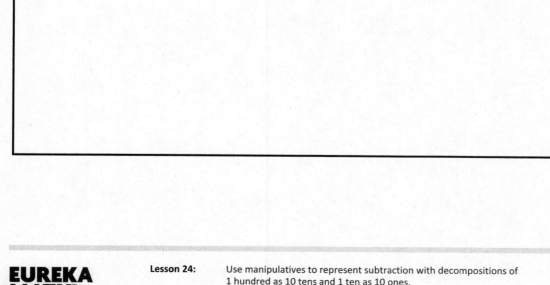

Lesson 24:   Use manipulatives to represent subtraction with decompositions of 1 hundred as 10 tens and 1 ten as 10 ones.

143

©2018 Great Minds®. eureka-math.org

W (Write a statement that matches the story.)

_____

_____

_____

Lesson 24: Use manipulatives to represent subtraction with decompositions of 1 hundred as 10 tens and 1 ten as 10 ones.

EUREKA MATH

Name _____ Date _____

1. Solve using mental math. If you cannot solve mentally, use your place value chart and place value disks.

 a. 25 – 5 = __20__ 25 – 6 = __19__ 125 – 25 = __100__ 125 – 26 = __99__

 b. 160 – 50 = __110__ 160 – 60 = __100__ 160 – 70 = __100__

2. Solve using your place value chart and place value disks. Unbundle the hundred or ten when necessary. Circle what you did to model each problem.

a. 124 – 60 = _____	b. 174 – 58 = _____
I unbundled the hundred. Yes No I unbundled a ten. Yes No	I unbundled the hundred. Yes No I unbundled a ten. Yes No
c. 121 – 48 = _____	d. 125 – 67 = _____
I unbundled the hundred. Yes No I unbundled a ten. Yes No	I unbundled the hundred. Yes No I unbundled a ten. Yes No
e. 145 – 76 = _____	f. 181 – 72 = _____
I unbundled the hundred. Yes No I unbundled a ten. Yes No	I unbundled the hundred. Yes No I unbundled a ten. Yes No

  Lesson 24: Use manipulatives to represent subtraction with decompositions of 1 hundred as 10 tens and 1 ten as 10 ones. 145

©2018 Great Minds®. eureka-math.org

g. 111 − 99 = _____ I unbundled the hundred. Yes No I unbundled a ten. Yes No	h. 131 − 42 = _____ I unbundled the hundred. Yes No I unbundled a ten. Yes No
i. 123 − 65 = _____ I unbundled the hundred. Yes No I unbundled a ten. Yes No	j. 132 − 56 = _____ I unbundled the hundred. Yes No I unbundled a ten. Yes No
k. 145 − 37 = _108_ I unbundled the hundred. Yes (No) I unbundled a ten. (Yes) No	l. 115 − 48 = _____ I unbundled the hundred. Yes No I unbundled a ten. Yes No

3. There were 167 apples. The students ate 89 apples. How many apples were left?

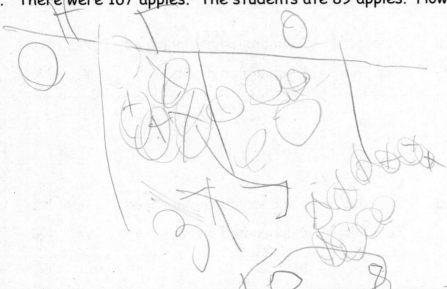

Lesson 24:   Use manipulatives to represent subtraction with decompositions of
1 hundred as 10 tens and 1 ten as 10 ones.

©2018 Great Minds®. eureka-math.org

EUREKA MATH™

For early finishers:

4. Tim and John have 175 trading cards together. John has 88 cards.

   a. How many cards does Tim have?

   b. Brady has 29 fewer cards than Tim. Have many cards does Brady have?

Lesson 24:   Use manipulatives to represent subtraction with decompositions of
                1 hundred as 10 tens and 1 ten as 10 ones.

©2018 Great Minds®. eureka-math.org

Name _____   Date _____

Solve using your place value chart and place value disks. Change 1 hundred for 10 tens and change 1 ten for 10 ones when necessary. Circle what you need to do to model each problem.

1.	2.
157 – 74 = _____	124 – 46 = _____
I unbundled the hundred. Yes No I unbundled a ten. Yes No	I unbundled the hundred. Yes No I unbundled a ten. Yes No

R (Read the problem carefully.)

114 people went to the fair. 89 of them went in the evening. How many went during the day?

D (Draw a picture.)

W (Write and solve an equation.)

W (Write a statement that matches the story.)

_____

_____

_____

EUREKA MATH™

Name _____ Date _____

1. Solve the following problems using the vertical form, your place value chart, and place value disks. Unbundle a ten or hundred when necessary. Show your work for each problem.

a. 72 – 49	b. 83 – 49
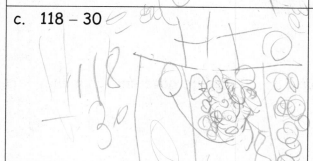	

c. 118 – 30	d. 118 – 85

e. 145 – 54	f. 167 – 78
	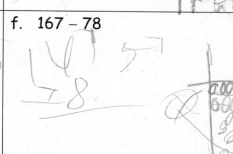

g. 125 – 87	h. 115 – 86
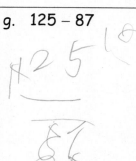	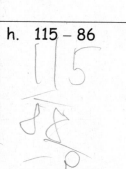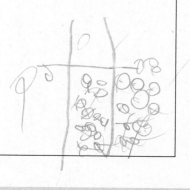

2. Mrs. Tosh baked 160 cookies for the bake sale. She sold 78 of them. How many cookies does she have left?

$160 - 78 =$

$100 - 78 =$

3. Tammy had $154. She bought a watch for $86. Does she have enough money left over to buy a $67 bracelet?

Lesson 25: Relate manipulative representations to a written method.

©2018 Great Minds®. eureka-math.org

EUREKA MATH

Name _____   Date _____

Solve the following problems using the vertical form, your place value chart, and place value disks. Unbundle a ten or hundred when necessary. Show your work for each problem.

1. 97 – 69

2. 121 – 65

R (Read the problem carefully.)

Chloe needs 153 beads to make a bag. She only has 49. How many more beads does she need?

D (Draw a picture.)

W (Write and solve an equation.)

Lesson 26:   Use math drawings to represent subtraction with up to two
               decompositions and relate drawings to a written method.

©2018 Great Minds®. eureka-math.org

157

W (Write a statement that matches the story.)

_____

_____

_____

Lesson 26:   Use math drawings to represent subtraction with up to two decompositions and relate drawings to a written method.

EUREKA MATH™

Name _____ Date _____

1. Solve vertically. Draw chips on the place value chart. Unbundle when needed.

   a. 181 – 63 = _____

hundreds	tens	ones

   b. 134 – 52 = _____

hundreds	tens	ones

   c. 175 – 79 = _____

hundreds	tens	ones

EUREKA
MATH™

Lesson 26:   Use math drawings to represent subtraction with up to two
decompositions and relate drawings to a written method.

©2018 Great Minds®. eureka-math.org

159

d. 115 − 26 = _____

hundreds	tens	ones

e. 110 − 74 = _____

hundreds	tens	ones

2. Tanisha and James drew models on their place value charts to solve this problem: 102 − 47. Tell whose model is incorrect and why.

James   Tanisha

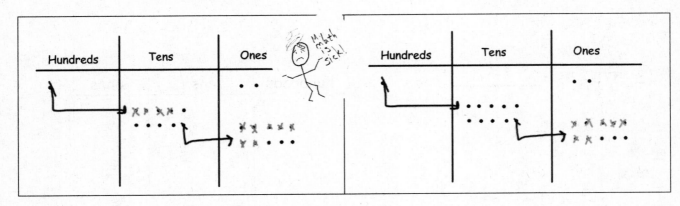

_____'s model is incorrect because _____

_____.

EUREKA MATH

Name _____ Date _____

Solve vertically. Draw chips on the place value chart. Unbundle when needed.

1. 153 – 46 = _____

hundreds	tens	ones

2. 118 – 79 = _____

hundreds	tens	ones

EUREKA MATH

Lesson 26: Use math drawings to represent subtraction with up to two decompositions and relate drawings to a written method.

©2018 Great Minds®. eureka-math.org

161

R (Read the problem carefully.)

Mr. Ramos has 139 pencils and 88 erasers. How many more pencils than erasers does he have?

D (Draw a picture.)

W (Write and solve an equation.)

W (Write a statement that matches the story.)

_____

_____

_____

Lesson 27:   Subtract from 200 and from numbers with zeros in the tens place.

EUREKA MATH

Name _____   Date _____

1. Make each equation true.

    a.   1 hundred = _____ tens

    b.   1 hundred = 9 tens _____ ones

    c.   2 hundreds = 1 hundred _____ tens

    d.   2 hundreds = 1 hundred 9 tens _____ ones

2. Solve vertically. Draw chips on the place value chart. Unbundle when needed.

    a.   100 – 61 = _____

hundreds	tens	ones

    b.   100 – 79 = _____

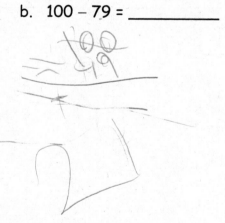

hundreds	tens	ones

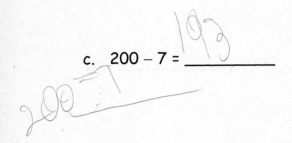

c.   200 – 7 = __193__

hundreds	tens	ones

d.   200 – 87 = __113__

hundreds	tens	ones

e.   200 – 126 = _____

hundreds	tens	ones

Lesson 27:   Subtract from 200 and from numbers with zeros in the tens place.

©2018 Great Minds®. eureka-math.org

EUREKA MATH™

Name _____ Date _____

Solve vertically. Draw chips on the place value chart. Unbundle when needed.

1. 100 – 44 = _____

hundreds	tens	ones

2. 200 – 76 = _____

hundreds	tens	ones

EUREKA MATH

Lesson 27:   Subtract from 200 and from numbers with zeros in the tens place.

©2018 Great Minds®. eureka-math.org

167

R (Read the problem carefully.)

Jerry made 200 pizzas. He sold some of them and had 57 pizzas left.

How many did he sell?

D (Draw a picture.)

W (Write and solve an equation.)

Lesson 28:   Subtract from 200 and from numbers with zeros in the tens place.

©2018 Great Minds®. eureka-math.org

169

W (Write a statement that matches the story.)

_____

_____

_____

Lesson 28:   Subtract from 200 and from numbers with zeros in the tens place.

EUREKA
MATH™

Name _____   Date _____

1. Solve vertically. Draw chips on the place value chart. Unbundle when needed.

a.   109 – 56 =

hundreds	tens	ones

b.   103 – 34 = ___68___

hundreds	tens	ones

c.   200 – 155 =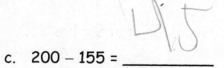

hundreds	tens	ones

EUREKA MATH

Lesson 28:   Subtract from 200 and from numbers with zeros in the tens place.

©2018 Great Minds®. eureka-math.org

171

d. 200 – 123 = _____

hundreds	tens	ones

2. Solve vertically without a place value chart.

200 – 148 = _____

3. Solve vertically. Draw a place value chart and chips.

Ralph has 137 fewer stamps than his older brother. His older brother has 200 stamps. How many stamps does Ralph have?

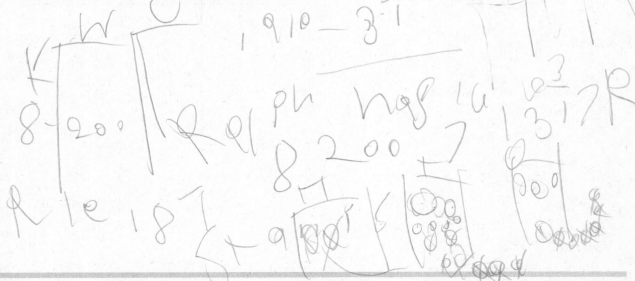

Lesson 28:   Subtract from 200 and from numbers with zeros in the tens place.

©2018 Great Minds®. eureka-math.org

EUREKA
MATH

Name _____ Date _____

Solve vertically. Draw chips on the place value chart. Unbundle when needed.

1. 108 – 79 = _____

hundreds	tens	ones

2. 200 – 126 = _____

hundreds	tens	ones

Lesson 28:   Subtract from 200 and from numbers with zeros in the tens place.

©2018 Great Minds®. eureka-math.org

173

R (Read the problem carefully.)

Kathy read 15 fewer pages than Lucy. Lucy read 51 pages. How many pages did Kathy read?

D (Draw a picture.)

W (Write and solve an equation.)

Lesson 29:   Use and explain the totals below method using words, math drawings, and numbers.

©2018 Great Minds®. eureka-math.org

175

W (Write a statement that matches the story.)

_____

_____

_____

Lesson 29:   Use and explain the totals below method using words, math drawings, and numbers.

©2018 Great Minds®. eureka-math.org

EUREKA
MATH™

Name _____   Date _____

1. Solve each addition expression using both the totals below and new groups below methods. Draw a place value chart with chips and two different number bonds to represent each.

   a. 27 + 19

New Groups Below	Totals Below	Place Value Chart	Number Bonds

   b. 57 + 36

New Groups Below	Totals Below	Place Value Chart	Number Bonds

EUREKA MATH

Lesson 29:   Use and explain the totals below method using words, math drawings, and numbers.

©2018 Great Minds®. eureka-math.org

177

2. Add like units and record the totals below.

a.
$$\begin{array}{r} 8\,7 \\ +\ 9\,5 \\ \hline \end{array}$$

_12_ (7 + 5) *ones*

_17°_ (80 + 90) *tens*

[182] 70 100

b.
$$\begin{array}{r} 1\,0\,6 \\ +\ \ \ 2\,4 \\ \hline \end{array}$$

10
20
100

[130]

c.
$$\begin{array}{r} 1\,5\,1 \\ +\ \ \ 4\,5 \\ \hline \end{array}$$

6
90
100

[196]

d.
$$\begin{array}{r} 1\,2\,6 \\ +\ \ \ 7\,2 \\ \hline \end{array}$$

e.
$$\begin{array}{r} 1\,5\,9 \\ +\ \ \ 3\,0 \\ \hline \end{array}$$

9   9+0

f.
$$\begin{array}{r} 1\,0\,8 \\ +\ \ \ 9\,1 \\ \hline \end{array}$$

Lesson 29:   Use and explain the totals below method using words, math drawings, and numbers.

©2018 Great Minds®. eureka-math.org

EUREKA MATH

Name _____   Date _____

Add like units and record the totals below.

1.   45
  + 64
―――
―――
▢
	2. 109
	  + 72
	―――
	―――
	―――
	▢

3.   144
  + 58
―――
―――
―――
▢
	4. 167
	  + 52
	―――
	―――
	―――
	▢

EUREKA MATH™

Lesson 29:   Use and explain the totals below method using words, math drawings, and numbers.

©2018 Great Minds®. eureka-math.org

179

R (Read the problem carefully.)

Eli spent 87 cents for a notebook and 38 cents for a pencil. How much money did he spend in all?

D (Draw a picture.)

W (Write and solve an equation.)

Lesson 30:   Compare totals below to new groups below as written methods.

©2018 Great Minds®. eureka-math.org

181

W (Write a statement that matches the story.)

_____

_____

_____

EUREKA
MATH

Name _____   Date _____

1. Linda and Keith added 127 + 59 differently. Explain why Linda's work and Keith's work are both correct.

Linda's work:	Keith's work:
$\begin{array}{r} 127 \\ +\ 59 \\ \hline 16 \\ 70 \\ +100 \\ \hline 186 \end{array}$	$\begin{array}{r} 127 \\ +\ 59 \\ \hline 186 \end{array}$

_____

_____

_____

2. Jake solved 124 + 69 using new groups below. Solve the same problem another way.

$\begin{array}{r} 124 \\ +\ 69 \\ \hline 193 \end{array}$	

3. Solve each problem two different ways.

a. 134 + 48	b. 83 + 69
c. 46 + 75	d. 63 + 128

EUREKA
MATH™

Name _____   Date _____

1. Kevin solved 166 + 25 using totals below. Solve the same problem another way.

| 166
+ 25
———
11
80
100
———
191 | |

2. Explain how Kevin's work and your work are similar.

_____

_____

_____

EUREKA MATH™   **Lesson 30:**   Compare totals below to new groups below as written methods.   185

©2018 Great Minds®. eureka-math.org

Name _____   Date _____

Solve the following word problems by drawing a tape diagram. Use any strategy you have learned to solve.

1. Mr. Roberts graded 57 tests on Friday and 43 tests on Saturday. How many tests did Mr. Roberts grade?

2. There are 54 women and 17 fewer men than women on a boat.

   a. How many men are on the boat?

   b. How many people are on the boat?

3. Mark collected 27 fewer coins than Craig. Mark collected 58 coins.

   a. How many coins did Craig collect?

   b. Mark collected 18 more coins than Shawn. How many coins did Shawn collect?

4. There were 35 apples on the table. 17 of the apples were rotten and were thrown out. 9 apples were eaten. How many apples are still on the table?

Lesson 31:   Solve two-step word problems within 100.

©2018 Great Minds®. eureka-math.org

EUREKA MATH

Name _____   Date _____

Solve the following word problems by drawing a tape diagram. Then, use any strategy that you've learned to solve.

1. Sandra has 46 fewer coins than Martha. Sandra has 57 coins.

   a. How many coins does Martha have?

   b. How many coins do Sandra and Martha have together?

2. There are 32 brown dogs and 19 white dogs at the park. 16 more brown dogs come to the park. How many dogs are there now at the park?

Grade 2
Module 5

R (Read the problem carefully.)

The shelter rescued 27 kittens in June. In July, 11 kittens were rescued.

In August, 40 more were rescued.

    a. How many kittens did the shelter rescue during those 3 months?

    b. If 64 of those kittens found homes by the end of August, how many still needed homes?

D (Draw a picture.)

W (Write and solve an equation.)

EUREKA
MATH™

Lesson 1:   Relate 10 more, 10 less, 100 more, and 100 less to addition and subtraction of 10 and 100.

193

©2018 Great Minds®. eureka-math.org

W (Write a statement that matches the story.)

a. _____

_____

b. _____

_____

Lesson 1:   Relate 10 more, 10 less, 100 more, and 100 less to addition and
subtraction of 10 and 100.

©2018 Great Minds®. eureka-math.org

EUREKA
MATH™

Name _____   Date _____

1. Complete each *more* or *less* statement.

    a. 10 more than 175 is _____.

    b. 100 more than 175 is _____.

    c. 10 less than 175 is _____.

    d. 100 less than 175 is _____.

    e. 319 is 10 more than _____.

    f. 499 is 100 less than _____.

    g. _____ is 100 less than 888.

    h. _____ is 10 more than 493.

    i. 898 is _____ than 998.

    j. 607 is _____ than 597.

    k. 10 more than 309 is _____.

    l. 309 is _____ than 319.

2. Complete each regular number pattern.

    a. 170, 180, 190, _____, _____, _____

    b. 420, 410, 400, _____, _____, _____

    c. 789, 689, _____, _____, _____, 289

    d. 565, 575, _____, _____, _____, 615

    e. 724, _____, _____, _____, 684, 674

    f. _____, _____, _____, 886, 876, 866

EUREKA MATH™   Lesson 1:   Relate 10 more, 10 less, 100 more, and 100 less to addition and   195
                              subtraction of 10 and 100.

©2018 Great Minds®. eureka-math.org

3. Complete each statement.

   a. $389 \xrightarrow{+10}$ _____ $\xrightarrow{+100}$ _____

   b. $187 \xrightarrow{-100}$ _____ $\xrightarrow{-10}$ _____

   c. $609 \xrightarrow{-10}$ _____ $\xrightarrow{-\,\underline{}} 499 \xrightarrow{+10}$ _____ $\xrightarrow{+\,\underline{}} 519$

   d. $512 \xrightarrow{-10}$ _____ $\xrightarrow{-10}$ _____ $\xrightarrow{+100}$ _____ $\xrightarrow{+100}$ _____ $\xrightarrow{+10}$ _____

4. Solve using the arrow way.

   a. 210 + 130 = _____

   b. 320 + _____ = 400

   c. _____ + 515 = 735

Lesson 1: Relate 10 more, 10 less, 100 more, and 100 less to addition and subtraction of 10 and 100.

©2018 Great Minds®. eureka-math.org

EUREKA MATH

Name _____ Date _____

Solve using the arrow way.

1. 440 + 220 = _____

2. 670 + _____ = 890

3. _____ + 765 = 945

Lesson 1: Relate 10 more, 10 less, 100 more, and 100 less to addition and
subtraction of 10 and 100.

197

ones	tens	hundreds

hundreds place value chart

Lesson 1:   Relate 10 more, 10 less, 100 more, and 100 less to addition and subtraction of 10 and 100.

©2018 Great Minds®. eureka-math.org

EUREKA MATH™

unlabeled hundreds place value chart

Lesson 1:   Relate 10 more, 10 less, 100 more, and 100 less to addition and
            subtraction of 10 and 100.

©2018 Great Minds®. eureka-math.org

201

R (Read the problem carefully.)

Max has 42 marbles in his marble bag after he added 20 marbles at noon.

How many marbles did he have before noon?

D (Draw a picture.)

W (Write and solve an equation.)

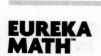

Lesson 2:   Add and subtract multiples of 100, including counting on to subtract.

203

©2018 Great Minds®. eureka-math.org

W (Write a statement that matches the story.)

_____

_____

_____

Lesson 2:   Add and subtract multiples of 100, including counting on to subtract.

EUREKA MATH

Name _____ Date _____

1. Solve each addition problem using place value strategies. Use the arrow way or mental math, and record your answers. You may use scrap paper if you like.

   a. 2 hundreds 4 tens + 3 hundreds = _____ hundreds _____ tens

      240 + 300 = _____

   b. 340 + 300 = _____   140 + 500 = _____   200 + 440 = _____

   c. 400 + 374 = _____   274 + 500 = _____   700 + 236 = _____

   d. 571 + _____ = 871   _____ + 349 = 749   96 + _____ = 696

   e. _____ + 562 = 862   300 + _____ = 783   600 + _____ = 726

2. Solve each subtraction problem using place value strategies. Use the arrow way or mental math, and record your answers. You may use scrap paper if you like.

   a. 6 hundreds 2 ones – 4 hundreds = _____ hundreds _____ tens _____ ones

      602 – 400 = _____

   b. 640 – 200 = _____   650 – 300 = _____   750 – _____ = 350

   c. 462 – 200 = _____   667 – 500 = _____   731 – 400 = _____

   d. 431 – _____ = 131   985 – _____ = 585   768 – _____ = 68

   e. _____ – 200 = 662   _____ – 300 = 653   734 – _____ =234

3. Fill in the blanks to make true number sentences. Use place value strategies, number bonds, or the arrow way to solve.

   a. 200 more than 389 is _____.

   b. 300 more than _____ is 568.

   c. 400 less than 867 is _____.

   d. _____ less than 962 is 262.

4. Jessica's lemon tree had 526 lemons. She gave away 300 lemons. How many does she have left? Use the arrow way to solve.

EUREKA MATH

Name _____   Date _____

Solve using place value strategies. Use the arrow way or mental math, and record your answers. You may use scrap paper if you like.

1. 760 – 500 = _____   880 – 600 = _____   990 – _____ = 590

2. 534 – 334 = _____   _____ – 500 = 356   736 – _____ = 136

R (Read the problem carefully.)

A children's library sold 27 donated books. Now, they have 48. How many books were there to begin with?

D (Draw a picture.)

W (Write and solve an equation.)

W (Write a statement that matches the story.)

_____

_____

_____

Lesson 3:   Add multiples of 100 and some tens within 1,000.

Name _____ Date _____

1. Solve each set of problems using the arrow way.

a.
380 + 200
380 + 220
380 + 230

b.
470 + 400
470 + 430
470 + 450

c.
650 + 200
650 + 250
650 + 280

d.
430 + 300
430 + 370
430 + 390

2. Solve using the arrow way or mental math. Use scrap paper if needed.

a. 490 + 200 = _____	210 + 490 = _____	490 + 220 = _____
b. 230 + 700 = _____	230 + 710 = _____	730 + 230 = _____
c. 260 + 240 = _____	260 + 260 = _____	280 + 260 = _____
d. 160 + 150 = _____	370 + 280 = _____	380 + 450 = _____
e. 430 + 290 = _____	660 + 180 = _____	370 + 270 = _____

3. Solve.

a. 66 tens + 20 tens = _____ tens   b. 66 tens + 24 tens = _____ tens

c. 66 tens + 27 tens = _____ tens   d. 67 tens + 28 tens = _____ tens

e. What is the value of 86 tens? _____

EUREKA MATH

Name _____   Date _____

Solve each set of problems using the arrow way.

1.

      440 + 300

      360 + 440

      440 + 380

2.

      670 + 230

      680 + 240

      250 + 660

R (Read the problem carefully.)

Diane needs 65 craft sticks to make a gift box. She only has 48. How many more crafts does she need?

D (Draw a picture.)

W (Write and solve an equation.)

Lesson 4: Subtract multiples of 100 and some tens within 1,000.

215

W (Write a statement that matches the story.)

_____

_____

_____

Name _____   Date _____

1. Solve using the arrow way.

a.
570 – 200
570 – 270
570 – 290

b.
760 – 400
760 – 460
760 – 480

c.
950 – 500
950 – 550
950 – 580

d.
820 – 320
820 – 360
820 – 390

EUREKA MATH™

2. Solve using the arrow way or mental math. Use scrap paper if needed.

a.
    530 – 400 = _____   530 – 430 = _____   530 – 460 = _____

b.
    950 – 550 = _____   950 – 660 = _____   950 – 680 = _____

c.
    640 – 240 = _____   640 – 250 = _____   640 – 290 = _____

d.
    740 – 440 = _____   740 – 650 = _____   740 – 690 = _____

3. Solve.

a. 88 tens – 20 tens = _____   b. 88 tens – 28 tens = _____

c. 88 tens – 29 tens = _____   d. 84 tens – 28 tens = _____

e. What is the value of 60 tens? _____

f. What is the value of 56 tens? _____

Lesson 4:   Subtract multiples of 100 and some tens within 1,000.

EUREKA MATH

Name _____   Date _____

1. Solve using a simplifying strategy. Show your work if needed.

    830 – 530 = _____   830 – 750 = _____   830 – 780 = _____

2. Solve.

    a. 67 tens – 30 tens = _____ tens. The value is _____.

    b. 67 tens – 37 tens = _____ tens. The value is _____.

    c. 67 tens – 39 tens = _____ tens. The value is _____.

R (Read the problem carefully.)

Jenny had 39 collectible cards in her collection. Tammy gave her 36 more. How many collectible cards does Jenny have now?

D (Draw a picture.)

W (Write and solve an equation.)

W (Write a statement that matches the story.)

_____

_____

_____

Lesson 5: Use the associative property to make a hundred in one addend.

EUREKA MATH

Name _____   Date _____

1. Solve.

    a. 30 tens = _____   b. 43 tens = _____

    c. 18 tens + 12 tens = _____ tens   d. 18 tens + 13 tens = _____ tens

    e. 24 tens + 19 tens = _____ tens   f. 25 tens + 29 tens = _____ tens

2. Add by drawing a number bond to make a hundred. Write the simplified equation and solve.

    a. 190 + 130

        10   120

    _____200 + 120_____ = _____

    b. 260 + 190

    _____ = _____

    c. 330 + 180

    _____ = _____

d. 440 + 280

_____ = _____

e. 199 + 86

_____ = _____

f. 298 + 57

_____ = _____

g. 425 + 397

_____ = _____

Lesson 5:   Use the associative property to make a hundred in one addend.

EUREKA
MATH

Name _____   Date _____

1. Add by drawing a number bond to make a hundred. Write the simplified equation and solve.

   a. 390 + 210

       _____ = _____

   b. 798 + 57

       _____ = _____

2. Solve.

   53 tens + 38 tens = _____

R (Read the problem carefully.)

Maria made 60 cupcakes for the school bake sale. She sold 28 cupcakes on the first day. How many cupcakes did she have left?

D (Draw a picture.)

W (Write and solve an equation.)

Lesson 6:   Use the associative property to subtract from three-digit numbers and verify solutions with addition.

227

©2018 Great Minds®. eureka-math.org

W (Write a statement that matches the story.)

_____

_____

_____

Lesson 6:   Use the associative property to subtract from three-digit numbers and verify solutions with addition.

©2018 Great Minds®. eureka-math.org

Name _____   Date _____

1. Draw and label a tape diagram to show how to simplify the problem. Write the new equation , and then subtract.

   a. 220 − 190 = ___230 − 200___ = _____

+ 10	220
+ 10	190

   b. 320 − 190 = _____ = _____

   c. 400 − 280 = _____ = _____

   d. 470 − 280 = _____ = _____

   e. 530 − 270 = _____ = _____

EUREKA
MATH™

Lesson 6:   Use the associative property to subtract from three-digit numbers and
verify solutions with addition.

©2018 Great Minds®. eureka-math.org

229

2. Draw and label a tape diagram to show how to simplify the problem. Write a new equation, and then subtract. Check your work using addition.

a. 451 – 199 = ____452 – 200____ = _____

	Check:
+1 | 451 +1 | 199	

b. 562 – 299 = _____ = _____

	Check:

c. 432 – 298 = _____ = _____

	Check:

d. 612 – 295 = _____ = _____

	Check:

Lesson 6:   Use the associative property to subtract from three-digit numbers and verify solutions with addition.

EUREKA MATH

Name _____ Date _____

Draw and label a tape diagram to show how to simplify the problem. Write the new equation, and then subtract.

1. 363 – 198 = _____ = _____

2. 671 – 399 = _____ = _____

3. 862 – 490 = _____ = _____

Lesson 6:   Use the associative property to subtract from three-digit numbers and
            verify solutions with addition.

231

©2018 Great Minds®. eureka-math.org

R (Read the problem carefully.)

Jeannie got a pedometer to count her steps. The first hour, she walked

43 steps. The next hour, she walked 48 steps.

   a. How many steps did she walk in the first two hours?

   b. How many more steps did she walk in the second hour than in

      the first?

D (Draw a picture.)

W (Write and solve an equation.)

Lesson 7:   Share and critique solution strategies for varied addition and
               subtraction problems within 1,000.

233

©2018 Great Minds®. eureka-math.org

W (Write a statement that matches the story.)

a. _____

_____

b. _____

_____

Lesson 7: Share and critique solution strategies for varied addition and subtraction problems within 1,000.

 EUREKA MATH™

Name _____   Date _____

1. Circle the student work that shows a *correct* solution to 543 + 290.

	Explain the mistake in any of the incorrect solutions.
$543 + 290 = 533 + 300 = 833$ $\wedge$ 533 10	
$543 + 290 = 553 + 300 = 853$ +10 [543] +10 [290]	_____ _____ _____
$543 \xrightarrow{+200} 743 \xrightarrow{+60} 803 \xrightarrow{+30} 833$	_____ _____

2. Circle the student work that *correctly* shows a strategy to solve 721 − 490.

$721 - 490 = 711 - 500 = 211$
711 $\wedge$ 10

+10 [721]
+10 [490]

$731 - 500 = 231$

Fix the work that is *incorrect* by making a new drawing in the space below with a matching number sentence.

EUREKA MATH   Lesson 7: Share and critique solution strategies for varied addition and subtraction problems within 1,000.   235

©2018 Great Minds®. eureka-math.org

3. Two students solved 636 + 294 using two different strategies.

$636 \xrightarrow{+4} 640 \xrightarrow{+60} 700 \xrightarrow{+30} 730 \xrightarrow{+200} 930$

$636 + 294 = 630 + 300 = 930$

630　6

Explain which strategy would be easier to use when solving and why.

_____

_____

_____

4. Circle one of the strategies below, and use the circled strategy to solve 290 + 374.

a.	b. Solve:
arrow way / number bond	

c. Explain why you chose that strategy.

_____

_____

_____

Lesson 7:　Share and critique solution strategies for varied addition and subtraction problems within 1,000.

©2018 Great Minds®. eureka-math.org

EUREKA MATH

Name _____   Date _____

Circle one of the strategies below, and use the circled strategy to solve 490 + 463.

a.	b. Solve:
arrow way / number bond	

c. Explain why you chose that strategy.

_____

_____

_____

EUREKA MATH™   Lesson 7:   Share and critique solution strategies for varied addition and   237
                                    subtraction problems within 1,000.

©2018 Great Minds®. eureka-math.org

Student A	Student B

Student A

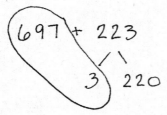

697 + 223

$$700 + 220 = 920$$

Student B

$$697 \xrightarrow{+3} 700 \xrightarrow{+200} 900 \xrightarrow{+20} 920$$

Student C

864 − 380

844   20

$$844 - 400 = 444$$

Student D

+20	864

+20	380

$$884 - 400 = 484$$

student work samples

R (Read the problem carefully.)

Susan has 37 pennies. M.J. has 55 more pennies than Susan.

   a. How many pennies does M.J. have?

   b. How many pennies do they have altogether?

D (Draw a picture.)

W (Write and solve an equation.)

W (Write a statement that matches the story.)

a. _____

_____

b. _____

_____

Lesson 8: Relate manipulative representations to the addition algorithm.

EUREKA MATH™

Name _____   Date _____

1. Solve the following problems using your place value chart, place value disks, and vertical form. Bundle a ten or hundred, when necessary.

a. 301 + 49	b. 402 + 48
c. 315 + 93	d. 216 + 192
e. 545 + 346	f. 565 + 226
g. 222 + 687	h. 164 + 745

2. Solve.

a. 300 + 200 = _____

b. 320 + 200 = _____

c. 320 + 230 = _____

d. 320 + 280 = _____

e. 328 + 286 = _____

f. 600 + 80 = _____

g. 600 + 180 = _____

h. 620 + 180 = _____

i. 680 + 220 = _____

j. 680 + 230 = _____

Lesson 8: Relate manipulative representations to the addition algorithm.

EUREKA MATH™

Name _____   Date _____

Solve the following problems using your place value chart, place value disks, and vertical form. Bundle a ten or hundred, when necessary.

1. 378 + 113

2. 178 + 141

R (Read the problem carefully.)

The table represents the halftime score at a basketball game. The red team scored 19 points in the second half. The yellow team scored 13 points in the second half.

Team	Score
red team	63 points
yellow team	71 points

a. Who won the game?

b. By how much did that team win?

D (Draw a picture.)

W (Write and solve an equation.)

W (Write a statement that matches the story.)

a. _____

_____

b. _____

_____

Lesson 9:   Relate manipulative representations to the addition algorithm.

EUREKA MATH

Name _____ Date _____

1. Solve the following problems using place value disks, a place value chart, and vertical form.

a. 417 + 293	b. 526 + 185
c. 338 + 273	d. 625 + 186
e. 250 + 530	f. 243 + 537
g. 376 + 624	h. 283 + 657

2. Solve.

a. 270 + 430 = _____

b. 260 + 440 = _____

c. 255 + 445 = _____

d. 258 + 443 = _____

e. 408 + 303 = _____

f. 478 + 303 = _____

g. 478 + 323 = _____

Lesson 9:   Relate manipulative representations to the addition algorithm.

EUREKA
MATH™

Name _____ Date _____

Solve the following problems using your place value chart, place value disks, and vertical form. Bundle a ten or hundred, when necessary.

1. 375 + 197

2. 184 + 338

R (Read the problem carefully.)

Benjie has 36 crayons. Ana has 12 fewer crayons than Benjie.

    a. How many crayons does Ana have?

    b. How many crayons do they have altogether?

D (Draw a picture.)

W (Write and solve an equation.)

Lesson 10:   Use math drawings to represent additions with up to two
compositions and relate drawings to the addition algorithm.

253

©2018 Great Minds®. eureka-math.org

W (Write a statement that matches the story.)

a. _____

_____

b. _____

_____

Lesson 10:   Use math drawings to represent additions with up to two
compositions and relate drawings to the addition algorithm.

EUREKA MATH

Name _____   Date _____

1. Solve using vertical form, and draw chips on the place value chart. Bundle as needed.

hundreds	tens	ones

a. 117 + 170 = _____

hundreds	tens	ones

b. 217 + 173 = _____

hundreds	tens	ones

c. 371 + 133 = _____

EUREKA MATH

Lesson 10:   Use math drawings to represent additions with up to two
compositions and relate drawings to the addition algorithm.

©2018 Great Minds®. eureka-math.org

255

hundreds	tens	ones

d. 504 + 269 = _____

2. Solve using vertical form, and draw chips on a place value chart. Bundle as needed.

a. 546 + 192 = _____

b. 546 + 275 = _____

Lesson 10:   Use math drawings to represent additions with up to two
compositions and relate drawings to the addition algorithm.

EUREKA
MATH

Name _____   Date _____

Solve using vertical form, and draw chips on a place value chart. Bundle as needed.

1. 436 + 509 = _____

2. 584 + 361 = _____

EUREKA MATH

Lesson 10:   Use math drawings to represent additions with up to two
compositions and relate drawings to the addition algorithm.

257

©2018 Great Minds®. eureka-math.org

R (Read the problem carefully.)

Mr. Arnold has a box of pencils. He passes out 27 pencils and has 45 left.

How many pencils did Mr. Arnold have in the beginning?

D (Draw a picture.)

W (Write and solve an equation.)

Lesson 11:   Use math drawings to represent additions with up to two
              compositions and relate drawings to the addition algorithm.

©2018 Great Minds®. eureka-math.org

259

W (Write a statement that matches the story.)

_____

_____

_____

Lesson 11:   Use math drawings to represent additions with up to two
compositions and relate drawings to the addition algorithm.

Name _____   Date _____

1. Solve using vertical form, and draw chips on the place value chart. Bundle as needed.

hundreds	tens	ones

a. 227 + 183 = _____

hundreds	tens	ones

b. 424 + 288 = _____

hundreds	tens	ones

c. 638 + 298 = _____

EUREKA MATH

Lesson 11:   Use math drawings to represent additions with up to two
compositions and relate drawings to the addition algorithm.

©2018 Great Minds®. eureka-math.org

261

hundreds	tens	ones

d. 648 + 289 = _____

2. Solve using vertical form, and draw chips on a place value chart. Bundle as needed.

a. 307 + 187

b. 398 + 207

Lesson 11: Use math drawings to represent additions with up to two
compositions and relate drawings to the addition algorithm.

©2018 Great Minds®. eureka-math.org

EUREKA MATH

Name _____   Date _____

Solve using vertical form, and draw chips on a place value chart. Bundle as needed.

1. 267 + 356 = _____

2. 623 + 279 = _____

Lesson 11:   Use math drawings to represent additions with up to two
                  compositions and relate drawings to the addition algorithm.

©2018 Great Minds®. eureka-math.org

263

Name _____   Date _____

1. Tracy solved the problem 299 + 399 four different ways.

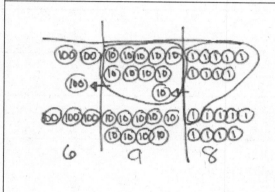

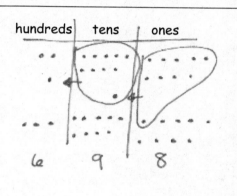

Explain which strategy is most efficient for Tracy to use and why.

_____

_____

_____

_____

_____

EUREKA
MATH™   **Lesson 12:**   Choose and explain solution strategies and record with a written   265
                                 addition method.

©2018 Great Minds®. eureka-math.org

2. Choose the best strategy and solve. Explain why you chose that strategy.

a. 221 + 498	Explanation: _____ _____ _____ _____
b. 467 + 200	Explanation: _____ _____ _____ _____
c. 378 + 464	Explanation: _____ _____ _____ _____

Lesson 12:   Choose and explain solution strategies and record with a written addition method.

EUREKA MATH

Name _____   Date _____

Choose the best strategy and solve. Explain why you chose that strategy.

1. 467 + 298	Explanation:
2. 300 + 524	Explanation:

EUREKA MATH™

Lesson 12:   Choose and explain solution strategies and record with a written addition method.

©2018 Great Minds®. eureka-math.org

267

R (Read the problem carefully.)

A fruit seller buys a carton of 90 apples. Finding that 18 of them are rotten, he throws them away. He sells 22 of the ones that are left on Monday. Now, how many apples does he have left to sell?

D (Draw a picture.)

W (Write and solve an equation.)

Lesson 13:   Relate manipulative representations to the subtraction algorithm, and
                   use addition to explain why the subtraction method works.

©2018 Great Minds®. eureka-math.org

269

W (Write a statement that matches the story.)

_____

_____

_____

Lesson 13:   Relate manipulative representations to the subtraction algorithm, and
use addition to explain why the subtraction method works.

Name _____   Date _____

1. Solve using mental math.

   a. 8 – 6 = _____   80 – 60 = _____   180 – 60 = _____   180 – 59 = _____

   b. 6 – 3 = _____   60 – 30 = _____   760 – 30 = _____   760 – 28 = _____

2. Solve using mental math or vertical form with place value disks. Check your work using addition.

   a. 138 – 17 = ___121___

         121
        + 17
         138

   b. 138 – 19 = _____

   c. 445 – 35 = _____

   d. 445 – 53 = _____

EUREKA MATH™

Lesson 13:   Relate manipulative representations to the subtraction algorithm, and use addition to explain why the subtraction method works.

271

©2018 Great Minds®. eureka-math.org

e. 863 – 170 = _____   f. 845 – 152 = _____

g. 472 – 228 = _____   h. 418 – 274 = _____

i. 567 – 184 = _____   j. 567 – 148 = _____

Lesson 13:   Relate manipulative representations to the subtraction algorithm, and
use addition to explain why the subtraction method works.

©2018 Great Minds®. eureka-math.org

EUREKA
MATH

Name _____   Date _____

Solve using mental math or vertical form with place value disks. Check your work using addition.

1. 378 – 117 = _____

2. 378 – 119 = _____

3. 853 – 433 = _____

4. 853 – 548 = _____

Lesson 13:   Relate manipulative representations to the subtraction algorithm, and
use addition to explain why the subtraction method works.

273

EUREKA MATH™

R (Read the problem carefully.)

Brienne has 23 fewer pennies than Alonzo. Alonzo has 45 pennies.

    a. How many pennies does Brienne have?

    b. How many pennies do Alonzo and Brienne have altogether?

D (Draw a picture.)

W (Write and solve an equation.)

Lesson 14:   Use math drawings to represent subtraction with up to two decompositions, relate drawings to the algorithm, and use addition to explain why the subtraction method works.

©2018 Great Minds®. eureka-math.org

275

W (Write a statement that matches the story.)

a. _____

_____

b. _____

_____

Lesson 14: Use math drawings to represent subtraction with up to two decompositions, relate drawings to the algorithm, and use addition to explain why the subtraction method works.

EUREKA MATH™

Name _____   Date _____

1. Solve by drawing place value disks on a chart. Then, use addition to check your work.

a. 469 – 170		Solve vertically or mentally:	Check:
b. 531 – 224		Solve vertically or mentally:	Check:
c. 618 – 229		Solve vertically or mentally:	Check:

Lesson 14:   Use math drawings to represent subtraction with up to two decompositions, relate drawings to the algorithm, and use addition to explain why the subtraction method works.

©2018 Great Minds®. eureka-math.org

277

d. 838 − 384	Solve vertically or mentally:	Check:
e. 927 − 628	Solve vertically or mentally:	Check:

2. If 561 − 387 = 174, then 174 + 387 = 561. Explain why this statement is true using numbers, pictures, or words.

Lesson 14: Use math drawings to represent subtraction with up to two decompositions, relate drawings to the algorithm, and use addition to explain why the subtraction method works.

EUREKA MATH™

Name _____ Date _____

Solve by drawing place value disks on a chart. Then, use addition to check your work.

1. 375 – 280	Solve vertically or mentally:	Check:
2. 741 – 448	Solve vertically or mentally:	Check:

Lesson 14: Use math drawings to represent subtraction with up to two decompositions, relate drawings to the algorithm, and use addition to explain why the subtraction method works.

©2018 Great Minds®. eureka-math.org

279

R (Read the problem carefully.)

Catriona earned 16 more stickers than Peter. She earned 35 stickers.

How many stickers did Peter earn?

MaryJo earned 47 stickers. How many more does Peter need to have the same amount as MaryJo?

D (Draw a picture.)

W (Write and solve an equation.)

Lesson 15: Use math drawings to represent subtraction with up to two decompositions, relate drawings to the algorithm, and use addition to explain why the subtraction method works.

281

©2018 Great Minds®. eureka-math.org

W (Write a statement that matches the story.)

_____

_____

_____

Lesson 15:   Use math drawings to represent subtraction with up to two decompositions, relate drawings to the algorithm, and use addition to explain why the subtraction method works.

©2018 Great Minds®. eureka-math.org

EUREKA
MATH™

Name _____ Date _____

1. Solve by drawing chips on the place value chart. Then, use addition to check your work.

a. 699 – 210 hundreds | tens | ones	Solve vertically or mentally:	Check:
b. 758 – 387 hundreds | tens | ones	Solve vertically or mentally:	Check:
c. 788 – 299 hundreds | tens | ones	Solve vertically or mentally:	Check:

Lesson 15:   Use math drawings to represent subtraction with up to two decompositions, relate drawings to the algorithm, and use addition to explain why the subtraction method works.

283

d. 821 – 523 hundreds tens ones	Solve vertically or mentally:	Check:
e. 913 – 558 hundreds tens ones	Solve vertically or mentally:	Check:

2. Complete all of the *if…then* statements. Draw a number bond to represent the related facts.

   a. If **762** – _____ = **173**, then **173 + 589** = _____.

   b. If **631** – _____ = **273**, then _____ + **273** = **631**.

Lesson 15:   Use math drawings to represent subtraction with up to two
                                    decompositions, relate drawings to the algorithm, and use addition to
                                    explain why the subtraction method works.

                     ©2018 Great Minds®. eureka-math.org

EUREKA MATH™

Name _____ Date _____

Solve by drawing chips on the place value chart. Then, use addition to check your work.

1. 583 – 327	Solve vertically or mentally:	Check:
hundreds	tens	ones		

2. 721 – 485	Solve vertically or mentally:	Check:
hundreds	tens	ones		

EUREKA
MATH™

Lesson 15: Use math drawings to represent subtraction with up to two
decompositions, relate drawings to the algorithm, and use addition to
explain why the subtraction method works.

©2018 Great Minds®. eureka-math.org

285

R (Read the problem carefully.)

Will read 15 more pages than Marcy. Marcy read 38 pages. The book is
82 pages long.

    a. How many pages did Will read?

    b. How many more pages does Will need to read to finish the book?

D (Draw a picture.)

W (Write and solve an equation.)

EUREKA MATH™

Lesson 16:   Subtract from multiples of 100 and from numbers with zero in the tens place.

©2018 Great Minds®. eureka-math.org

287

W (Write a statement that matches the story.)

a.
_____

_____

b.
_____

_____

Lesson 16: Subtract from multiples of 100 and from numbers with zero in the tens place.

©2018 Great Minds®. eureka-math.org

Name _____   Date _____

1. Solve vertically or using mental math. Draw chips on the place value chart and unbundle, if needed.

a. 304 – 53 = _____

hundreds	tens	ones

b. 406 – 187 = _____

hundreds	tens	ones

c. 501 – 316 = _____

hundreds	tens	ones

Lesson 16:   Subtract from multiples of 100 and from numbers with zero in the tens place.

289

d. 700 – 509 = _____

hundreds	tens	ones

e. 900 – 626 = _____

hundreds	tens	ones

2. Emily said that 400 – 247 is the same as 399 – 246. Write an explanation using pictures, numbers, or words to prove Emily is correct.

Lesson 16:   Subtract from multiples of 100 and from numbers with zero in the tens place.

EUREKA
MATH™

Name _____   Date _____

Solve vertically or using mental math. Draw chips on the place value chart and unbundle, if needed.

1. 604 – 143 = _____

hundreds	tens	ones

2. 700 – 568 = _____

hundreds	tens	ones

R (Read the problem carefully.)

Colleen put 27 fewer beads on her necklace than Jenny did. Colleen put on

46 beads. How many beads did Jenny put on her necklace?

If 16 beads fell off of Jenny's necklace, how many beads are still on it?

D (Draw a picture.)

W (Write and solve an equation.)

Lesson 17:   Subtract from multiples of 100 and from numbers with zero in the tens place.

©2018 Great Minds®. eureka-math.org

293

W (Write a statement that matches the story.)

_____

_____

_____

Lesson 17:   Subtract from multiples of 100 and from numbers with zero in the tens place.

©2018 Great Minds®. eureka-math.org

EUREKA
MATH™

Name _____   Date _____

1. Solve vertically or using mental math. Draw chips on the place value chart and unbundle, if needed.

   a. 200 – 113 = _____

hundreds	tens	ones

   b. 400 – 247 = _____

hundreds	tens	ones

   c. 700 – 428 = _____

hundreds	tens	ones

Lesson 17:   Subtract from multiples of 100 and from numbers with zero in the tens place.

d. 800 – 606 = _____

hundreds	tens	ones

e. 901 – 404 = _____

hundreds	tens	ones

2. Solve 600 – 367. Then, check your work using addition.

Solution:	Check:

Lesson 17:   Subtract from multiples of 100 and from numbers with zero in the tens place.

©2018 Great Minds®. eureka-math.org

EUREKA MATH™

Name _____   Date _____

Solve vertically or using mental math. Draw chips on the place value chart and unbundle, if needed.

1. 600 – 432 = _____

hundreds	tens	ones

2. 303 – 254 = _____

hundreds	tens	ones

R (Read the problem carefully.)

Joseph collected 49 golf balls from the course. He still had 38 fewer than his friend Ethan.

   a. How many golf balls did Ethan have?

   b. If Ethan gave Joseph 24 golf balls, who had more golf balls? How many more?

D (Draw a picture.)

W (Write and solve an equation.)

Lesson 18:   Apply and explain alternate methods for subtracting from multiples of
100 and from numbers with zero in the tens place.

©2018 Great Minds®. eureka-math.org

299

W (Write a statement that matches the story.)

a.
_____

_____

b.
_____

_____

Lesson 18:   Apply and explain alternate methods for subtracting from multiples of 100 and from numbers with zero in the tens place.

EUREKA MATH™

Name _____ Date _____

1. Use the arrow way and counting on to solve.

a. 300 – 247	b. 600 – 465

2. Solve vertically, and draw a place value chart and chips. Rename in one step.

a. 507 – 359	b. 708 – 529

3. Choose a strategy to solve, and explain why you chose that strategy.

a. 600 – 437	Explanation:

EUREKA MATH™

Lesson 18: Apply and explain alternate methods for subtracting from multiples of 100 and from numbers with zero in the tens place.

©2018 Great Minds®. eureka-math.org

301

b. 808 – 597	Explanation:

4. Prove the student's strategy by solving both problems to check that their solutions are the same. Explain to your partner why this way works.

$$\begin{array}{r} 800 \\ -543 \\ \hline \end{array} \qquad \begin{array}{r} 799 \\ -542 \\ \hline \end{array}$$

5. Use the simplifying strategy from Problem 4 to solve the following two problems.

a. 600 – 547	b. 700 – 513

Lesson 18:   Apply and explain alternate methods for subtracting from multiples of 100 and from numbers with zero in the tens place.

©2018 Great Minds®. eureka-math.org

EUREKA MATH

Name _____ Date _____

Choose a strategy to solve, and explain why you chose that strategy.

1. 400 – 265	Explanation:
2. 507 – 198	Explanation:

Lesson 18:   Apply and explain alternate methods for subtracting from multiples of
100 and from numbers with zero in the tens place.

©2018 Great Minds®. eureka-math.org

303

Name _____   Date _____

1. Explain how the two strategies to solve 500 – 211 are related.

a.

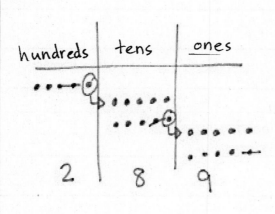

hundreds	tens	ones
2	8	9

b.

$$
\begin{array}{r}
5\cancel{0}\cancel{0} \\
-2\,1\,1 \\
\hline
2\,8\,9
\end{array}
$$

_____

_____

_____

_____

_____

_____

_____

EUREKA MATH

Lesson 19:   Choose and explain solution strategies and record with a written addition or subtraction method.

©2018 Great Minds®. eureka-math.org

305

2. Solve and explain why you chose that strategy.

a. 220 + 390 = _____	Explanation: _____ _____ _____ _____
b. 547 – 350 = _____	Explanation: _____ _____ _____ _____
c. 464 + 146 = _____	Explanation: _____ _____ _____ _____
d. 600 – 389 = _____	Explanation: _____ _____ _____ _____

Lesson 19:   Choose and explain solution strategies and record with a written addition or subtraction method.

EUREKA MATH™

Name _____   Date _____

Solve and explain why you chose that strategy.

1. 400 + 590 = _____	Explanation:
	_____
	_____
	_____
	_____
2. 775 − 497 = _____	Explanation:
	_____
	_____
	_____
	_____

EUREKA MATH™   **Lesson 19:**   Choose and explain solution strategies and record with a written addition or subtraction method.   307

©2018 Great Minds®. eureka-math.org

Name _____ Date _____

Step 1:   Show your strategy to solve.

Step 2:   Find a classmate who used a different strategy, and copy his work into the box.

Step 3:   Discuss which strategy is more efficient.

1.   399 + 237 = _____

a. My strategy	b. _____'s strategy

2.   400 – 298 = _____

a. My strategy	b. _____'s strategy

Lesson 20:   Choose and explain solution strategies and record with a written addition or subtraction method.

309

©2018 Great Minds®. eureka-math.org

3. 548 + 181 = _____

a. My strategy	b. _____'s strategy

4. 360 + _____ = 754

a. My strategy	b. _____'s strategy

5. 862 − _____ = 690

a. My strategy	b. _____'s strategy

Lesson 20: Choose and explain solution strategies and record with a written addition or subtraction method.

EUREKA MATH

Name _____ Date _____

Solve each problem using two different strategies.

1. 299 + 156 = _____

a. First Strategy	b. Second Strategy

2. 547 + _____ = 841

a. First Strategy	b. Second Strategy

Lesson 20:   Choose and explain solution strategies and record with a written
                 addition or subtraction method.

311

©2018 Great Minds®. eureka-math.org

Credits

Great Minds® has made every effort to obtain permission for the reprinting of all copyrighted material. If any owner of copyrighted material is not acknowledged herein, please contact Great Minds for proper acknowledgment in all future editions and reprints of this module.